Biodegradable Polymers

Materials and their Structures

Manjari Sharma
Assistant Professor
Department of Applied Sciences and Humanities
Ambedkar Institute of Advanced Communication Technologies & Research
(Formerly Ambedkar Institute of Technology)
Geeta Colony, East Delhi - 110031

CRC Press is an imprint of the
Taylor & Francis Group, an **informa** business

First published 2021
by CRC Press
2 Park Square, Milton Park, Abingdon, Oxon, OX14 4RN

and by CRC Press
6000 Broken Sound Parkway NW, Suite 300, Boca Raton, FL 33487-2742

© 2021 Manakin Press Pvt. Ltd.

CRC Press is an imprint of Informa UK Limited

The right of Manjari Sharma to be identified as author of this work has been asserted by him in accordance with sections 77 and 78 of the Copyright, Designs and Patents Act 1988.

Reasonable efforts have been made to publish reliable data and information, but the author and publisher cannot assume responsibility for the validity of all materials or the consequences of their use. The authors and publishers have attempted to trace the copyright holders of all material reproduced in this publication and apologize to copyright holders if permission to publish in this form has not been obtained. If any copyright material has not been acknowledged please write and let us know so we may rectify in any future reprint.

All rights reserved. No part of this book may be reprinted or reproduced or utilised in any form or by any electronic, mechanical, or other means, now known or hereafter invented, including photocopying and recording, or in any information storage or retrieval system, without permission in writing from the publishers.

For permission to photocopy or use material electronically from this work, access www.copyright.com or contact the Copyright Clearance Center, Inc. (CCC), 222 Rosewood Drive, Danvers, MA 01923, 978-750-8400. For works that are not available on CCC please contact mpkbookspermissions@tandf.co.uk

Trademark notice: Product or corporate names may be trademarks or registered trademarks, and are used only for identification and explanation without intent to infringe.

Print edition not for sale in South Asia (India, Sri Lanka, Nepal, Bangladesh, Pakistan or Bhutan).

British Library Cataloguing-in-Publication Data
A catalogue record for this book is available from the British Library

Library of Congress Cataloging-in-Publication Data
A catalog record has been requested

ISBN: 978-0-367-77476-9 (hbk)
ISBN: 978-1-003-17159-1 (ebk)

Dedicated

To my parents

Brief Contents

1. Introduction — 1–28

2. Classification of Biodegradable Polymers — 29–50

3. Materials and Their Structures — 51–86

4. Innovation Till Date — 87–136

5. Characterization Techniques — 137–174

6. Biodegradation — 175–208

Glossary — 209–216

Index — 217–220

Detailed Contents

1. Introduction   1–28
     1.0   Introduction   3
     1.1   Background   5
     1.2   Definitions   6
     1.3   Developmental Need   9
     1.4   Recycling   10
     1.5   Process Compatibilities   11
     1.6   Need of Development of Biodegradable Polymer   11
     1.7   Challenge in Present Scenario   11
     1.8   Manufacturers of Biodegradable Polymers as on Date   12
     1.9   Commercial Biodegradable Products of Starch   14
     1.10   Applications of Biodegradable Polymer   15
     1.11   Objectives for Development   19
     1.12   Current Status in India   20
     1.13   Current Status in World   23
     1.14   Conclusion   24
     1.15   References   25

2. Classification of Biodegradable Polymers   29–50
     2.0   Categories of Polymers   31
     2.1   Non-degradable and Non-renewable Polymers   32
     2.2   Degradable and Non-renewable Polymers   33
     2.3   Biodegradable and Non-renewable Polymers   34
     2.4   Biodegradable and Renewable Polymers   36

2.5	Categories of Biodegradable Polymers	37
2.6	Details of Different Categories Based Biodegradable Polymers	40
2.7	References	49

3. Materials and Their Structures 51–86

3.0	Biodegradable Polymer - Materials and their Structures	53
3.1	Polysaccharides	53
	3.1.1 Starch	53
	3.1.2 Cellulose	62
	3.1.3 Chitin and Chitosan	64
3.2	Protein- Based	65
3.3	Bacterial/Microbial Polyesters	67
3.4	Poly (Glycolic Acid)	68
3.5	Synthetic Polyester	68
	3.5.1 Polycaprolactone	68
	3.5.2 Poly (Lactic Acid)	69
3.6	Poly (Vinyl Alcohol)	72
3.7	Starch/Poly (Vinyl Alcohol) (PVA) Based Biodegradable Polymers	73
3.8	Starch/Poly (Epsilon-Caprolactone) (PCL) Based Biodegradable Polymers	74
3.9	Starch/Poly (D, L-Lactic Acid) (PLA) Based Biodegradable Polymers	75
3.10	Starch / Poly-hydroxy-alkanoates (PHA) Based Biodegradable Polymers	75
3.11	Starch/Poly (Butylenes Succinate-Co-Butylene Adipate) (PBSA) Based Biodegradable Polymers	78
3.12	Starch with Olefins Vinyl Derivatives	78
3.13	Starch and Olefins with Acrylate Derivatives	80
3.14	Starch and Olefins with Anhydride Group	80
3.15	References	81

4. Innovation Till Date   87–136

    4.0   Research in the Field of Biodegradable Polymers as on Date   89

    4.1   References   122

5. Characterization Techniques   137–174

    5.0   Characterization Techniques   139

    5.1   Physico-chemical Analysis   139

           5.1.1   FTIR: ATR Fourier Transform Infrared: Attenuated Total Reflectance: Spectroscopy   139

    5.2   Thermal Analysis   144

           5.2.1   DSC: Differential Scanning Calorimetery   144

           5.2.2   TGA: Thermo Gravimetric Analysis   151

           5.2.3   Thermo Mechancial Analyser (TMA)   155

           5.2.4   Dynamic Mechanical Analyser (DMA)   156

    5.3   Crystallographic Study   160

           5.3.1   XRD: X-Ray Diffraction   160

    5.4   Morphology Analysis   163

           5.4.1   SEM: Scanning Electron Microscopy   163

    5.5   Water Absorption Analysis   167

    5.6   Physico-mechanical Properties   167

           5.6.1   Tensile Strength   167

           5.6.2   Tear Strength   169

           5.6.3   Burst Strength   169

    5.7   Degradation Study   170

           5.7.1   Weight Loss Study   170

           5.7.2   Carbonyl Index Study   171

           5.7.3   Mechanical Properties After Degradation   171

           5.7.4   Morphology Study After Degradation   172

    5.8   References   172

6. Biodegradation 175–208

6.0	Biodegradation	177
6.1	Degradation Process of Bio-degradable Polymer	178
6.2	Factors Affecting Biodegradability	180
6.3	International Standard Methods for Bio-degradability Testing	181
6.4	Methods of Biodegradation	184
	6.4.1 Soil Burial	186
	6.4.2 Composting	186
6.5	Definition of Composting	188
	6.5.1 According to ASTM Standards	188
	6.5.2 According to ISO Draft Standards	188
	6.5.3 According to British Standards	188
	6.5.4 According to US Environmental Protection Agency (US EPA)	189
6.6	Factors Affecting Composting Process	189
6.7	Phases of Composting Process	190
6.8	An International Standard Related to Composting	193
	6.8.1 ASTM Standards Related to Composting	193
	6.8.2 ISO Standards Related to Composting	194
	6.8.3 EN Standard Related to Composting	195
6.9	Methods of Biodegradability Testing by Composting	196
	6.9.1 ISO 14855-1 and 2	196
6.10	Methods of Bio-degradation Testing	202
6.11	Biodegradation Supporting Environments	203
6.12	Duration of Bio-degradation and Risk of their Entry into Nature	204
6.13	Biodegradation Measurement Procedures	205
6.14	References	206

Glossary 209–216
Index 217–220

Preface

This book is about development of biodegradable polymers alternatives, which are required to save our reserves of fossil fuels and to save our mother earth from further environmental degradation. Petrochemical resources on earth are depleting day by day due to excessive usage. We use polymers in every aspect of our life and we have become highly dependent on them. In last two decades emphasis was on to develop stronger polymeric materials, which are unaffected even by the harsh environmental conditions, but now there is paradigm shift in approach i.e. Petrochemicals to Potato, because disposal of petrochemical based polymers is still an issue and something out of Potato as a substitute to Polymers is a nature to nature approach.

Research has shifted now to prepare biodegradable polymers, which possess the equal capabilities as of synthetic polymer, so that items prepared by this material remain useful throughout their useful life time and thereafter, they would biodegrade without much effort in a environmentally sound manner.

There is a race to find the best alternative having lowest cost and having more or less similar capacity as that of petrochemical resource based polymers. Alternatives are there but cost economics of sustainability does not permit us to adopt them. These alternatives have been derived from natural resources, which have been modified scientifically. These modifications are possible because of advanced technologies available & capabilities where it is possible to identify each resource, based on its property, modification and alternatives. Research continued so far has moved into a right direction without any doubt but it can further do miracles if it continues to grow a little faster. Actually the research should be result oriented rather than being for academic purpose.

There are many books in the field of polymers but there are very few books in this specialized field of developing biodegradable polymers and there is hardly any book in the field biodegradable polymer concentrating on development of biodegradable polymers with their characterization and providing help in conducting studies on prepared product's biodegradation through various instrumentation techniques.

This book is concerned with this hot topic to deal with the family of biodegradable polymers which have to be prepared with a novel idea of studying polymers with "Cradle to Grave" approach. It deals basic materials, which could be potential materials to prepare biodegradable polymers with their basic structures, properties, behavior and limitations known till date.

By referring to this book, a researcher can get a help in selection of material for the study of biodegradable polymer through finding collection of information for sources of polymer, availability of polymers, origin of polymers, nature of polymers, processing difficulties of polymers.

This book will help students in understanding various characterization techniques which can be used for the study of identification of functional group, structural properties, thermal behaviour, crystallographic nature, mechanical properties and morphological properties through FTIR–ATR for physico chemical properties, DSC and TGA for thermal studies, XRD for crystallographic studies and SEM for morphological studies. It also provided information regarding ASTM methods and procedures to characterize the samples through instrumentation.

I hope this book will be useful in developing ideas about biodegradable material and their testing procedures. It also provided an overview of various testing methods to analyze biodegradability including standard guideline for evaluation of biodegradation and compostability of polymer material through ASTM/ISO/EN standard methods.

It is difficult to acknowledge everyone who has helped me in preparation of this book. Many publications were motivating me to write a book concerning this area. I, sincerely would like to thank Prof. Anek Pal Gupta, Delhi Technological University, Prof. R. C. Sharma, Delhi Technological University, Dr. Vijai Kumar, CIPET, Bhopal, Sh. S. K. Saxena, CIPET, Ahmedabad, Dr. S. K. Shukla, Bhaskaracharya College of Applied Sciences, Dwarka, Delhi, Sh. R. K. Shukla, Delhi Technological University (Library), Delhi, Dr. Nivedita, Dr. Neerja Sahu for their valuable suggestions and support.

I am fortunate to have met in my professional career many persons who have helped me and inspired me each time. Special thanks is also due to my beloved husband Aditya Sharma, Scientist, Central Pollution Control Board, Delhi, and children Pranjal Sharma and Paridhi Sharma, my parents, my in-laws. Ms. Sharda Sharma, Dr. P. K. Sharma, Ms. Sunita Sharma, Dr. Vandana Sharma, my brothers Ravi, Sanjay, Gopal for their love and faith.

Manjari Sharma

List of Abbreviation

ABS	:	Acrylonitrile Butadiene Styrene
AN	:	Acrylonitrile
ASTM	:	American Society for Testing Materials
ATR	:	Attenuated Total Reflectance
BMP	:	Biochemical methane
BPS	:	Biodegradable Plastic Society Japan
CA	:	Cellulose Accetate
CAB	:	Cellulose acetate butyrate
CC	:	Chitosan composite
CEN	:	European Organization for Standardization
CMS	:	Carboxymethyl Starch
CS	:	Cationic Starch
DCP	:	Dicumyl Peroxide
DIC	:	Dissolved Inorganic Carbon
DIN	:	German Organization for Standardization
DMA	:	Dynamic Mechanical Analyser
DOC	:	Dissolved Organic Carbon
DS	:	Degree of Substitution

DSC	:	Differential Scanning Calorimeter
DTG	:	Derivative Thermograviemetry
EAA	:	Poly (ethyleneco-acrylicn acid)
EB	:	Electron Beam
EBP	:	Ethylene-1-butene Copolymer
EMMT	:	Ethanolamine-activated Montmorillonite
E-P	:	Ethylene-Propylene Copolymers
EPMA	:	Ethylene-Propylene-Maleic Anhydride Co-polymer
EVAL	:	Ethylene- Vinyl Alcohol Co-polymer
FETPS	:	Formamide-ethanolamine-plasticized Thermoplastic Starch
FTIR	:	Fourier Transform Infrared
GC	:	Glycerin Content
GL	:	Glycerol
GLU	:	Glutaraldehyde
GMS	:	Glycerol Monostearate
GTPS	:	Glycerol-plasticized Thermoplastic Starch
Hc	:	Heat of Crystallization
HCNs	:	Hemp Cellulose Nanocrystals
HDPE	:	High Density Polyethylene
Hf	:	Heat of Fusion
i-PP	:	Isotactic Polypropylene
ISO	:	International Organization for Standardization

List of Abbreviation

LDI	:	Lysine Disocynate
LDPE	:	Low Density Polyethylene
LLDPE	:	Linear Low-density Polyethylene
LVDT	:	Linear variable Displacement Transducer
MA	:	Malic Acid
MAA	:	Methacrylic Acid
mA-g	:	Maleic Anhydride Grafted
MAH	:	Maleic Anhydride
MAPP	:	Maleic Anhydride Grafted Polypropylene
MBS	:	Methyl Methacrylate - Butadiene-Styrene
MFI	:	Melt Flow Index
MMT	:	Montmorillonite
MSW	:	Municipal Solid Waste
MTPSC	:	Montmorillonite-reinforced Thermoplastic Starch Composites
NMR	:	Nuclar Magnetic Resonance
OCST	:	Starch (octanoated starch)
OECD	:	Organization for Economic Corporation and Development
PBAT	:	Poly (butylene Adipate terephthalate)
PBS	:	Poly (butylene Succinate)
PBSA	:	Poly (butylene Succinate Adipate)
PBS-Poly	:	Butylene Succinate
PBST	:	Poly (butylene Succinate Terephthalate)

PBT	:	Poly (butylene Terephthalate)
PCL	:	Poly (ε-Caprolactone)
PDLA	:	Poly (D-Lactide)
PDLLA	:	Poly-DL-lactide
PE	:	Polyethylene
PEA	:	Polyesteramides
PEG	:	Poly (ethylene glycol)
PE-g-MA	:	Polyethylene-g-maleic Anhydride
PET	:	Poly (Ethyleneterephthalate)
PGA	:	Poly (glycolic acid)
PHA	:	Poly Hydroxy Alkanoates
PHBV	:	Poly (3-Hydroxybutyrate-co-3 hydroxyvalarate)
PHEE	:	Poly (hydroxy ester ether)
PHV	:	Poly (3- Hydroxybutyrate)
PLA	:	Poly Lactic Acid
PLLA	:	Poly (L-Lactic Acid)
PMMA	:	Poly (methyl methacrylate)
POE	:	Polyethylene-octene-elastomer
PP	:	Polypropylene
PPC	:	Poly (propylene carbonate)
PPO	:	Polyphenylene oxide
PS	:	Polystyrene
PVA/PVOH	:	Poly (Vinyl Alcohol)

List of Abbreviation

PVAc	:	Poly Vinyl Accetate
PVC	:	Polyvinyl Chloride
RHF	:	Rice Husk Flour
SCA	:	Starch-copoymer-cellulose Acetate
SEM	:	Scanning Electron Microscopy
SEVA-C	:	Starch Copolymer-ethylene and Vinyl Alcohol
SMA	:	Styrene-Maleic Anhydride Co-polymer
SPCL	:	Starch-copolymer -poly-e-caprolactone
SPP	:	Sweet Potato Pulp
TA	:	Thermal Analysis
TA	:	Tartaric Acid
T_c	:	Crystallization Onset Temperature
T_g	:	Glass Transition Temperature
TGA	:	Thermo Gravimetric Analysis
Thbiogas	:	Theoritical Amount of Biogas Evolved
$ThCH_4$	:	Theoritical Amount of Evolved Methane
$ThCO_2$	:	Theoritical Amount of Evolved Carbondioxide
ThOD	:	Theoritical Oxygen Demand
TM	:	Melting Temperature
TMA	:	Thermo Mechanical Analyser
TMC	:	Total Moisture Contents
TO	:	Thermo – oxidized
TOC	:	Total Organic Carbon Content

TPF	:	Thermoplastic Wheat Flour
TPS	:	Thermo Plastic Starch
TS	:	Tensile Strength
WAXD	:	Wide Angle X-ray Diffraction
WF	:	Wood Flour
XPS	:	X-ray Photoelectron Spectroscopy
XRD	:	X-ray Diffraction

1
Introduction

INSIDE THIS CHAPTER

1.0 Introduction

1.1 Background

1.2 Definitions

1.3 Developmental Need

1.4 Recycling

1.5 Process Compatibilities

1.6 Need of Development of Biodegradable Polymer?

1.7 Challenge in Present Scenario?

1.8 Manufacturers of Biodegradable Polymers as on Date

1.9 Commercial Biodegradable Products of Starch

1.10 Applications of Biodegradable Polymer

1.11 Objectives for Development

1.12 Current Status in India

1.13 Current Status in World

1.14 Conclusion

1.15 References

1.0 INTRODUCTION

Petroleum resources are depleting day by day. Synthetic polymers produced from petroleum resources have taken a special space in our daily routine. This is due to their cost and easy fabrication process. They have been widely used by different sections of our society. In our daily life plastics are used to such an extent that the disposal of the used plastic products has created great problem. On the other hand there are certain applications where polymer needs limited life time *i.e.* biomedical application, disposal packaging, mulch films etc. Hence, there is a need to develop a biodegradable polymer which may be used in environmentally sound manner with respect to its specific use.

There is a shift in our approach from petroleum resource based processes to life sciences based processes. Natural materials has a broader base of materials, including plant derived products and fermentation products. The move from "petroleum to potatoes" is internationally driven. This is indeed a challenging field of research and innovation with unlimited future prospects.

If we carefully manage the entire process there is no chance of failure. Hence, we need to take care before choosing the products for our daily life use. Coming scenario in the world is approaching for the same. Although phrasing in words and converting into reality is challenging, but efforts made in the direction will definitely provide positive results.

The environmental and regulatory drivers are encouraging industry to develop and engineer products with an approach of "cradle to grave", which requires to manufacture the product from natural resources and to dispose-off the product in environmentally sound manner after their useful life.

At the same time societal drivers will have to take lead in accepting only environmentally friendly materials although they may cost little more in comparison. The societal change could be such a big factor which can have chain reaction and would compel the industry to only produce environmentally sound material even it may cost more. It can be seen in new generation, having been set by teaching community, that they are automatically tilted towards a better option, which is eco friendly irrespective of item cost.

Available resources can be classified as renewable and non-renewable. Renewable resources are generally those resources which can restock themselves at approximately the rate at which they have been extracted. Renewable resources like plants, forests, farm have many uses. There are three major plant based polymers. These plants are Protein, Oil and Carbohydrates. Polysaccharides are one of the predominant carbohydrates beside starch and cellulose. There are natural fibres falling in the carbohydrate family. Natural fibres such as hemp, straw, kenaf, jute, flax consists mainly of cellulose and lignin. Non-living renewable natural resources include water, wind, tides and solar radiation. In general renewable resources are totally natural resources that are not depleted when used by human beings. Plastics, gasoline, coal and other items produced from fossil fuels are non-renewable, because these resource are getting depleted day by day, and can be used only once.

The development of biodegradable polymer through natural materials is definitely a challenge to our scientific society, where entire responsibility of developing such polymer lies [1-3].

1.1 BACKGROUND

There are many natural polymers available in our surroundings like starch, cellulose, chitin, chitosen, protein based biopolymers, microbial polymers which can be used for different applications[4]. The agriculture industry produces sufficient supplies of some agricultural products that could be used as renewable resources for polymer feed stocks. These renewable resources are naturally biodegradable. These polymers are not designed for high temperature and therefore their uses are limited to normal ranges of biological function found in the biosphere.

Polymers from renewable resources offer an answer to maintain a sustainable development of economically and ecologically attractive technology. The advantages of the innovation in the development of new materials from renewable resources is likely to result into improved preservation of fossil-based raw materials, reduction in garbage volume and compostability in the natural environment. The possibilities of use of agricultural resources for the production of biodegradable materials have attracted great interest not only from academecia but also from industrial point of view.

First biodegradable synthetic polymer, the polyglycollic acid, was invented in 1954 and which has been used in general as a biodegradable suture in surgery. However, owing to poor thermal and hydrolytic stabilities, its applications have been limited. Off late, due to environmental consciousness, interest has arisen in modifying commodity polymers like polyethylene, polystyrene, polyacrylates, polyvinyl chloride etc. such that they also degrade partially or fully in natural environment.

Polysaccharides are the world's most abundant polymers as well as a renewable resource which are biodegradable and environmental friendly materials. Among polysaccharides the starch is one of the most important member. It's low cost and readily availability, almost in pure form, has picked the interest of researchers not only because it is a renewable raw material, but also for its potential to impart biodegradability to fabricated material. Starch and other biodegradable materials have been used as fillers in petroleum based plastics [5-11].

Packaging being one of the largest consuming sectors of plastics in the world, concern over the disposal of enormous tonnage of plastic packaging material has lead to the search of biodegradable plastics which can alleviate the problem of plastic waste. Griffin introduced an idea by adding starch to a synthetic polymer such as polyethylene by conventional compounding and processing. But the product obtained was paper like material.

1.2 DEFINITIONS

Several attempts have been made in the recent past to define terms such as "degradation" and "biodegradation" within environmental applications [12, 13]. The term "biodegradation" indicates a degradation process brought about by living organisms. Biodegradation resulting from environmental exposure, however, generally involves the action of micro-organisms, usually resulting in a reduction in degree of polymerization and degradation of the polymer to simple organic moieties.

1.2 Definitions

Technical and General definitions for "Biodegradable Plastic/Polymer" are:

(a) Technical definitions

Definition 1: Biodegradable plastics: Plastic material that disintegrates under environmental conditions in a reasonable and demonstrable period of time, where the primary mechanism is through micro-organisms such as bacteria, yeast, fungi and algae. Suggested definition developed at first International Scientific Consensus Workshop, Toronto, Canada, in November, 1989.

Definition 2: Biodegradable polymer is a polymer in which the degradation is mediated atleast partially by a biological system. The suggested working definition has been adopted by the second International Workshop on Biodegradable Polymers and Plastics, Montpellier, France, 1991.

Definition 3: Biodegradable Plastic: A degradable plastic in which the degradation results from the action of naturally occurring micro-organisms such as bacteria, fungi and algae. Definition has been suggested by International Standard Organization, ISO 472:1988.

Definition 4: Biodegradation: A process of decomposition facilitated by biochemical mechanisms. The definition is suggested at the International Workshop on Biodegradability, Annapolis, Maryland, 1992.

Definition 5: Biodegradable: Capable of being broken down into innocuous products by the action of living things.

The definition is suggested by Webster's Ninth New Collegiate Dictionary, 1989.

(b) General Definition

Biodegradation: A degradation facilitated by living organisms, usually micro-organisms [14]. Biodegradable polymers are defined as those that undergo microbially induced chain scission leading to mineralization. Thus starch is biodegradable because it can be readily metabolized by a wide array of organisms. During this bio-degradation process all carbon should be accounted for (carbon balance) and all residues should be non-toxic in the environmental assessment. In addition the residue and microbial biomass should eventually be incorporated into the natural geochemical cycle.

In 1992, an international workshop was organized to bring together experts from around the world to achieve areas of agreement on definitions, standards and testing methodologies. There is a general agreement concerning the following key points [15].

- For all practical purposes of applying a definition, material manufactured to be biodegradable must relate to a specific disposal pathway such as composting, sewage treatment, denitrification, anaerobic sludge treatment.

- The rate of degradation of a material manufactured to be biodegradable has to be consistent with the disposal method and other components of the pathway into which it is introduced, such that accumulation is controlled.

- The ultimate end products of aerobic biodegradation of a material manufactured to be biodegradable are carbon dioxide, water and minerals and that the intermediate products include biomass and humic materials.

- Material must biodegrade safely and not negatively impact the environment in any manner.

As a result standard organizations such as International Standards Organization (ISO) and American Society of Testing and Materials (ASTM) were encouraged to rapidly develop standards.

In 1994, ASTM updated these definitions. Biodegradable material is "capable of undergoing decomposition into carbon dioxides, methane, water, inorganic compounds or biomass in which the predominant mechanism is the enzymatic action of micro-organism that can be measured by standardized test in a specific period of time, reflecting available disposable condition" [16].

1.3 DEVELOPMENTAL NEED

Plastics are strong, light weight, inexpensive, easily process-able and energy efficient. Plastics are resistant to microbial attack because of their hydrophobic character and high molecular weight. Until the useful life time of plastics, these are desired properties but after their lifetime these properties become drawback because plastics do not break down to become the part of biological carbon cycle of our ecosystem. This resulted into an irreversible buildup of plastic material on the earth, in the environment, causing scarring of landscapes, causing blockage of waste water pipelines, fouling of beaches, and a serious hazard to marine life. It is reported that average consumption in Europe is nearly 100 kg per year per person[17].

1.4 RECYCLING

Recycling is one of the option to fight back the problem of disposal of plastics in environmental sound manner with a limitation that reclaimed polymer from waste is not a pure single polymer. It is a mixture of similar and or different polymers. It may consist of homopolymers, copolymers, polymer blends as well as a mixture of chain and step polymers, and /or thermoplastics and thermosets. This limitation is due to the difficulty of identification and separation of polymers after their useful life once they are thrown as garbage and mixed up with Municipal Solid Waste. To manufacture a specific property product mixing occurs at the manufacturing stage itself due to which segregation of specific plastics is very difficult. As a result, the reclaimed polymers through recycling are non homogeneous and consequently have non-uniform properties. Thermosets, if present in the mixture with thermoplastics in the waste, do not melt and create problem in size reduction operation by extruder. The properties of recycled plastics are generally inferior to that of virgin polymers in terms of mechanical strength, clarity, brightness etc. Reclaimed plastics can not be used as containers for food or pharmaceutical products because of their lower purity.

1.5 PROCESS COMPATIBILITIES

To conserve the environmental degradation it is essential to have compatibility with the processes, products and technologies associated. The waste streams generated from the manufacturing processes should also be recycled or converted into usable by-products instead of generating wastes. It will ensure minimum environmental impact, when such products are being made available for use by the manufacturing industries.

1.6 NEED OF DEVELOPMENT OF BIODEGRADABLE POLYMER?

- Firstly, the petroleum products are depleting very fast. As per recent survey the stocks shall last up to maximum 70 years. Hence, there is a direct need of replacement of these products by the natural/renewable products.

- Secondly, the petroleum based products used for the short term applications are attracting environmental concerns because:

  – if these materials are burnt, emit lots of emissions,

  – if dumped, do not get converted into 100% biomass and residues are occupying the precious earth space.

  – if left openly, they are harmful especially to marine life.

1.7 CHALLENGE IN PRESENT SCENARIO?

The challenge today is to develop the technology needed to make the bio based materials resolution a reality. More precisely we are looking for:

- Cost effective alternative to petroleum based products without compromising the properties.

- Material, which shows structural and functional stability during storage and use.

- Material susceptible to microbial and environmental biodegradation only upon disposal and without significant environmental impacts.

1.8 MANUFACTURERS OF BIODEGRADABLE POLYMERS AS ON DATE

Today, commercial biodegradable polymers in the open market are being offered by an increasing number of manufacturers. Although there are many different biodegradable polymers available, they mostly fall into one of the following groups:

- Polymers based on starch (starch based plastics);
- Polymers based on polylactic acid (polylactide, polylactic acid, PLA);
- Polymers based on polyhydroxyalkanoates (PHA's: PHB, PHBV, etc.);
- Polymers based on aliphatic-aromatic polyesters;
- Polymers based on cellulose (cellophane, etc.);
- Polymers based on lignin.

World's largest companies have shifted their approach from traditional petrochemical based technologies to life sciences based technologies. A list of major manufacturers of biodegradable polymers is given in Table 1.1.

1.8 Manufacturers of Biodegradable Polymers as On Date

Table 1.1: Major Manufacturers of Biodegradable Polymers

Sr. No.	Manufacturers	Biodegradable Polymers	Degradation
01	Novamont, Italy	Starch Based	Biodegradation
02	Union Carbide, Solvay	Poly-caprolactone	Biodegradation
03	Mitsui Toatsu, Cargill	Polylactic acid	Hydrolysis and Biodegradation
04	National Starch, USA	Starch foam	Biodegradation
05	Air products and Chemicals, USA	Polyvinyl alcohol	Biodegradation
06	Hoechst, Germany	Polyvinyl alcohol	Biodegradation
07	Ecostar, USA	Starch activator	Photo or Biodegradation
08	Dow Chemicals, USA	Polyethylene/CO	Photo degradation
09	Amko Plastics, USA	Polybioethylene	–
10	Warner-Lambert Company, USA	Starch Based	Biodegradation
11	Earthsoul, Mumbai	PCL and Starch based bio-products	Biodegradation
12	BASF	Polyester and Starch (Ecoflexresistant to water and grease suitable for hygienic disposable wrappings)	Biodegradation
13	Environmental Polymers (Woolston, Warrington, UK)	polyvinyl alcohol (Depart–for extrusion, injection and blow moulding having user controlled solubility in water for hospital laundry bags, disposable food items)	Biodegradation
14	Procter and Gamble	Nodax PBHB similar to ecoflex. (for transport of biohazard material)	Dissolved in high pH
15	Nature Tec Products (product of Harita NTI), Chennai	Importer of ecoflex, Novamount etc.	Biodegradation

(Contd...)

16	Surya Polymers, New Delhi	Importer of ecoflex, Novamount etc.	Biodegradation
17	Symphony Polymers, New Delhi	Importer of ecoflex	Biodegradation
18	Green Craft Polymers, New Delhi	Importer of ecoflex	Biodegradation
19	Overseas Polymers, New Delhi	Importer of ecoflex	Biodegradation
20	Sachdeva Plastics, New Delhi	Importer of ecoflex	Biodegradation

1.9 COMMERCIAL BIODEGRADABLE PRODUCTS OF STARCH

The plasticized starch alone is mainly used in soluble compostable foams, such as loose-fillers, expanded trays, shape moulded parts and expanded layers, as a replacement for polystyrene. BIOTEC of Germany has conducted promising research and development along the lines of starch-based materials. The company's three product lines are Bioplastm granules for injection moulding, Bioflexm film, and Biopurm foamed starch.

Under the Mater-Bi trademark, Novamont of Italy today produces four classes of biodegradable materials Z, Y, V, and A, all containing starch and differing in synthetic components. Each class is available in several grades and has been developed to meet the needs of specific applications. The current production capacity of Novamont is 8000 tons/year. Mater-Bi can be processed using conventional plastic technologies such as injection moulding, blow moulding, film blowing, foaming, thermoforming and extrusion. The physical-mechanical properties of Mater-Bi are similar to those of conventional plastics like polyethylene and polystyrene. Mater-Bi is not only recyclable but also as biodegradable as pure cellulose. The biodegradability of Mater-Bi products has been measured according to standard test methods approved by International Organizations (ISO, CEN and ASTM). The compostability of some Mater-Bi grades has been certified by the "Ok Compost" label. Mater-Bi can be used in a wide range of applications such as disposable items (plates, cutlery, cup lids etc.), packaging (wrapping film, film for dry food packaging, board lamination etc.), stationery (pens, cartridges, pencil sharpeners etc.), personal care and hygiene (sanitary napkins, soluble cotton swabs etc.) and a lot others like toys, shopping bags, mulch film etc.

Various starch plastics with different trade names are now available in the market. Buna Sow Leuna of Germany has developed a line of biodegradable polymers based on esterified starch with the trade names Sconacell S, Sconacell A, and Sconacell AF.

Compared to common thermoplastics, however, biodegradable products based on starch still reveal many disadvantages which are mainly attributed to the highly hydrophilic character of starch polymers. Inspite of many positive results, thermoplastic starch-based materials are still at an early stage of development and the markets for such products are expected to increase in future as the properties are more improved, prices still decline, and an infrastructure for composting becomes more established.

1.10 APPLICATIONS OF BIODEGRADABLE POLYMER

Biodegradable polymers are now produced to replace non-biodegradable plastics in several applications. The design of these materials generally begins with a conceptual application. It may be required to replace an existing material, or to complement one.

Single use plastics products are targeted as the primary market area. Some common sectors where applications for biopolymers have introduced include medicine, packaging, agriculture, and automobile industry. Many materials that have been developed and commercialized are applied in more than one of these categories.

Biopolymers deployed for packaging continue to receive more attention than those designated for any other application. All levels of government, particularly in China and Germany [18] are endorsing the widespread application of biodegradable packaging materials in order to reduce the volume of inert materials currently being disposed of in landfill sites, occupying scarce available space. It is estimated that 41% of plastics are used in packaging, and that almost half of that volume is used to package food products.

Biodegradable foams, films and molded articles for use as disposable plates and cutlery, shopping, compost bags, molded containers, films and sheets for packaging and mulch films have been introduced during the last few years. Some of the market areas for the biodegradable polymers are given in Table 1.2.

Table 1.2: Market Areas for the Biodegradable Polymers

S. No.	Area of use	Product form	Application
01	Agriculture	Fibre	Netting
		Film	Mulching, controlled release of agrochemicals.
02	Horticulture	Film	Nursery bags, packaging of perishable foods, dairy products, fruits, vegetables, flowers, hosiery etc.
		Moulded products	Plant covers, wind shield and plant holders.
03	Packaging	Blister packaging sheets and bubble films	Packaging of fragile goods.

(Contd...)

1.10 Applications of Biodegradable Polymer

04	Domestic	Films	Shopping bags, composting bags, diapers, feminine hygiene, Products, garbage bags.
		Moulded products	Food Containers, vegetable and fruit crates, egg-boxes, food service items, toys, pens, cutlery and cups, razors, containers for mineral water, soft drinks, beverages etc.
05	Hospital	Moulded products and Sheets, bags	Disposable needles and syringes, sutures, surgical gowns, blood bags, dextrose solution, bottles, hospital bed sheets, medicine trays, bags etc.

Agricultural applications for biopolymers are not limited to film covers. Containers such as biodegradable plant pots and disposable composting containers and bags are areas of interest. The pots are seeded directly into the soil, and breakdown as the plant begins to grow.

Fertilizer and chemical storage bags which are biodegradable are also applications that material scientists have examined. From an agricultural standpoint, biopolymers which are compostable are important, as they may supplement the current nutrient cycle in the soils where the remnants are added.

The medical world is constantly changing, and consequently the materials employed by it also see recurrent adjustments. The biopolymers used in medical applications must be compatible with the tissue they are found in, and may or may not be expected to break down after a given time period.

Researchers working in tissue engineering are attempting to develop organs from polymeric materials, which are fit for transplantation into humans. The plastics would require injections with growth factors in order to encourage cell and blood vessel growth in the new organ. Work completed in this area includes the development of biopolymers with adhesion sites that act as cell hosts in giving shapes that mimic different organs. Not all biopolymer applications in the field of medicine are as involved as artificial organs. The umbrella classification of bioactive materials includes all biopolymers used for medical applications. One example is artificial bone material which adheres and integrates onto bone in the human body.

The most commonly employed substance in this area is called Bioglass [19]. Another application for biopolymers is in controlled release delivery of medications. The bioactive material releases medication at a rate determined by its enzymatic degradation [20] PLA materials were developed for medical devices such as resorbable screws, sutures, and pins [21]. These materials reduce the risk of tissue reactions to the devices, shorten recovery times, and decrease the number of doctor visits needed by patients.

The automotive sector is responding to societal and governmental demands for environmental responsibility. Biobased cars are lighter, making them a more economical choice for consumers, as fuel costs are reduced. Natural fibres are substituted for glass fibres as reinforcement materials in plastic parts of automobiles and commercial vehicles. An additional advantage of using biodegradable polymer materials is that waste products may be composted. Natural fibres (from flax or hemp) are usually applied in formed interior parts. The components do not need load bearing capacities, but dimensional stability is important. Research and development in this area continues to be enthusiastic, especially in European countries.

There are a number of novel applications for biopolymers, which do not fit into any of the previous categories. One such example is the use of biopolymer systems to modify food textures. For example, biopolymer starch (elatine-based) fat replacers possess fat-like characteristics of smooth, short plastic textures that remain highly viscous after melting. Research continues into high pressure being used to manipulate biopolymers into food products. The eventual goal is improved physical characteristics such as foaming, gelling, and water- or fat-binding abilities. Biopolymer materials are currently incorporated into adhesives, paints, engine lubricants, and construction materials. Biodegradable golf tees and fishing hooks have also been invented. The attraction of biopolymers in all of these areas is their derivation from renewable sources, slowing the depletion of limited fossil fuel stores.

1.11 OBJECTIVES FOR DEVELOPMENT

The Challenge is to develop the technology needed to make the bio-based materials resolution — a reality.

Objectives as on date are to develop:

1. Cost effective alternative to petroleum based polymers.

2. Eco friendly material having structural and functional stability during storage and use.

3. Material, susceptible to microbial and environmental degradation after disposal without significant environmental impacts.

4. Since the most promising approach as on date seems to be: to synthesize the biodegradable polymer by compounding of natural polymer with synthetic polymer to achieve biodegradable polymer with this combination, the aims are:

- To obtain better compatibility and miscibility by the chemical reactivity.
- To improve the interaction at molecular level.
- To introduce abundant oxygen through incorporation of oxygen group, this shall help in degradation of polymer produced.
- To maintain the properties of the polymer similar to that of synthetic polymer used so that the properties of product like tensile strength, percentage elongation, least affinity with moisture etc. are better.
- To maintain the processability of synthetic polymer product.
- To develop biodegradable polymer ready to use as packaging material, or applicable for other purposes as biodegradable material, which may be used for commercial applications.

1.12 CURRENT STATUS IN INDIA

India generates 5.6 million metric tons of plastic waste annually, with Delhi generating the most of at municipality at 689.5 metric tons every day, according to a report from the Central Pollution Control Board (CPCB). About 60 percent of the total 9,205 metric tons per day is recycled. CPCB was asked to study plastic waste generation by India's

Supreme Court, which is continuing to discourage the sale of gutka – a popular stimulant consisting of crushed betel nut, chewing tobacco and other flavourings – in plastic pouches. CPCB submitted the report to the Indian Supreme Court, which said, "We are sitting on a plastic time bomb." The court asked authorities in five cities – Delhi, Agra, Jaipur, Faridabad and Bangalore – "to submit reports on the steps taken to contain dumping of plastic waste and implementing the ban on gutka."

The issues of biodegradable polymers are quite different in India, which are primarily due to the following reasons:

(a) **Cost:** The cost of biodegradable plastics is 2 – 10 times more than conventional polymers in India. Based on available information, the current price trend of Oxo/Photo-degradable polymers (Based on Polyethylene material) and Biodegradable polymers for film applications (co-polyester based) are:

   (*i*) Oxo/Photo Degradable polymers film/bags – ₹ 90–120 per kg (depending upon prices of polyethylene and additive, depending on the global pricing trend.

   (*ii*) Biodegradable polymers film / bags – ₹ 400 – 500 per kg.

(b) **Dearth of Initiatives:** There is hardly any legal framework to enforce legislature to acknowledge the disposal problem of conventional polymers, particularly for short lived flexible packaging products. Agro-biotech may be the new word for India's science and technology sector, but alternative biodegradable polymers have still not been seen as major area for research and innovation in the country. It is noticed that for export of products to some other countries, there is a mandatory condition for packaging that it should be made up of Biodegradable Polymers. Hence, there is a thought and it is being tried on a very small scale. In fact it has not been seen as a potential area.

(c) **Use of biodegradable polymers bags:** In India per capita consumption of polymer is quite low as compared to developed countries, because of Indian culture of using the items upto the last possible use of item before throwing in garbage. Even the small items like tooth brush are being used for many purposes before disposal in a life span of not less than a year. However, there is a steep increase in consumption in comparison. Use of biodegradable polymers will overall increase India's polymer consumption and it is opined that the major cause of concern would be to segregate between non-biodegradable and biodegradable polymers in the waste stream of Municipal Solid Waste (MSW). Hence, introduction of biodegradable polymers would be helpful only if proper "Eco-labelling" system and "Polymers Coding System" as given in Recycled Polymers Manufacture and Usage Rules, 1999, as amended in 2003 is implemented. An integrated waste management system has to be planned in order to effectively use, recycle and dispose of bio-polymeric materials[22]. It is important to mention that all levels of government, particularly in China [23] and Germany [18], have endorsed the widespread application of biodegradable materials used for packaging applications in order to reduce the volume of inert materials disposed of landfills, scarcity of space.

1.13 CURRENT STATUS IN WORLD

In USA during past two decades municipal solid waste generation has increased significantly. It has gone up to 236.2 millions tons per year which is nearly 50% high in comparison to its level of 1980. MSW contains largest component of organic matter, paper and paper board products upto 35% of the waste and food scarps and others near about 24%. Plastic comprises 11% *i.e.* 26.7 million tons at third place in MSW waste composition. Nearly 9% plastic containers have been recycled in USA.[24]

In European Union (EU) waste policy is based on waste hierarchy concept. It means that ideally, waste should be prevented and what can not be prevented should be reused, recycled and recovered as much as is feasible with landfill being used as little as possible. From 1995 to 2004 MSW in EU has constantly grown by about 2% per year from 204 million tonnes in 1995 to 243 million tonnes in 2003. EU citizen generates Municipal Waste, at an average of 550 kg/year.[25]

In Singapore, it is estimated that MSW generation is about 4.5 to 4.8 million tonnes per year and plastic accounts for 5.8% of the total soild waste.

In Australia, total recycling rate of plastic has increased from 7 to 12.4%. It is reported that plastics packaging recycling in 2003 was 134905 tonnes which is 20.5% of packaging consumption during a year.

In China, production of MSW is increasing rapidly. Average amount of MSW produced by each person daily has increased from 1.12 to 1.59 kg from 1986 to 1995. It is estimated that amount of MSW produced annually is about 204.40 to 440 kg. Total Solid waste produced in China is about 27.15% of that in Asia and 15.07% of that Globally[26].

1.14 CONCLUSION

There is huge potential and large number of areas where biodegradable polymer materials may find their usage. The sectors like agriculture, automotives, medicine, and packaging all require environmentally friendly polymers. Because the level of biodegradation may be customised to specific needs, each industry may create its own ideal material. The different modes of biodegradation are also advantageous for these kinds of materials, because disposal methods can be tailored as per industry requirements. Environmental responsibility is consistently on a sharp rise to both consumers and industry. For those who produce biodegradable plastic materials, there lies a key advantage. Biopolymers limit carbon dioxide emissions during creation, and degrade to organic matter after disposal easily in an environmentally sound manner which is one of the most important criterion. Although synthetic polymers are economically more feasible because of established technologies as compared to biodegradable polymers, an increased availability of biodegradable plastics will compel many consumers to choose them on the basis of their environmentally sound disposal mechanism[26]. The processes which hold the most promising approach for further development of biopolymer materials are[27], which use renewable resources. Biodegradable polymers containing starch and/or cellulose fibres appear to be the most likely option in war future expecting continual growth.

Microbially grown polymers seem to be scientifically sound, but the infrastructure required to commercially expand their use is still very costly, and difficult to create. Time has come for the exponential growth of biodegradable polymer development, as the society awareness is now compelling manufacturers to develop such materials.

1.15 REFERENCES

[1] G. Griffin (Ed.), Chemistry and Technology of Biodegradable Polymers, *Chapman and Hall*, 1994.

[2] R.W. Lenz, *Adv Polym Sci*, 107;1: 1993.

[3] M. R. Timmins and R. W. Lenz, *Trends Polym*, 2;18:1994.

[4] High Performance Fibers, Ed. J.W.S. Hearle, *Woodhead Publishing Limited*, 2001.

[5] G.F. Fanta, C.L. Swanson, and R. L. Shogren, *J Appl Polym Sci*, 44; 2037:1992.

[6] S. T. Lian, J. L. Jane, S. Rajagopalan, and P. A. Seib, *Biotechnol Prog*, 8; 51:1992.

[7] R. L. Evangelista, Z. L. Nikoldu, W. Sang., J. Jane, and R. L. Gelina, *Ind. Eng. Chem. Res*, 30; 1841:1991.

[8] F. H. Otey and R.P. Westoff. *U.S. Pat.* 4,133,784 (1979).

[9] G.J.L. Griffin., *Adv. Chem. Ser.*, 134; 159:1974.

[10] R. P. Westoff., F. H. Otey, C. L. Mehltreften, and C. R. Russel, *Ind. Eng. Chem. Prod, Res. Dev*, 13; 123:1974.

[11] E. H. Park, E. R. George, M. A. Muldoon, and A. Flammini, *Polym, News*, 19(8); 230–238:1996.

[12] S.A. Barenberg, J.L. Brash, R. Narayan, and A. E. Redpath (eds.), Degradable Materials: Perspectives, Issues and opportunities, CRC Press, Boca Raton, Florida, 1990.

[13] V. Coma, Y. Couturier, B. Pascat, G. Bureau, S. Guilbert, and J. L. Cuq, in Biodegradable Polymers and Plastics (M. Vert, J. Feijen, A. Albertsson, G. Scott, and E. Chellini, Eds.), Roya Soc Chem., Cambridge, England, p-242: 1992.

[14] Anthony L. Andrady, JMS-Rev. Macromol Chem. Phys, C34(1); 25–76: 1994.

[15] Towards Common Ground-Meeting Summary of the International Workshop on Biodegradability, Annapolis, MD, USA, 1992.

[16] Z. Z. Dechev, Chapter 27, Handbook of Engineering Biopolymers, S. Fakirov and D. Bhattacharyya, Hanser Publication, Munich Germany, p-800: 2007.

[17] K.F Mulder, Technological Forecasting and Social Change, 58; 105–124: 1998.

[18] C. Bastioli, *Macromolecular Symposia*, 135(1); 193–204:1998.

[19] T. Kokubo, HM Kim and M. Kawashita, Biomaterials, 24(13); 2161–75: 2003.

[20] S. E. Sakiyama-Elbert and JA Hubbell, *Materials Research*, 31; 183–201:2001.

[21] W. M. Raymond and L.K. Selin, *Nature Immunology* , 2; 417–426: 2002.

[22] P. M. Subramanian, Resources, Conservation, and Recycling., 28; 253–263: 2000.

[23] H Chau., Yu P., *Water Science and T echnology*, 39(10–11); 273–280:1999.

1.15 References

[24] Municipal Solid Waste Generation, Recycling, and Disposal in United States; Facts and Figures: 2003, United States Environmental Protection Agency.

[25] Watste Generated and Treated in Europe. Data 19995; 2003 European Commission, Eurostat, Luxembourg, 2005.

[26] Wei. Y. S., Fan Y. V. Wang M. J., Wang, J.S.; Composting and Compost Application in China, Resourc: Conserve. Recycl. 30; 277: 2002

[26] E. Grigat, R Kock, R. Timmermann, Polymer Degradation and Stability. 59(1–3): 223–226: 1998.

[27] AK. Ashwin, K. Karthick, and K.P. Arumugam, *International Journal of Chemical Engineering and Applications*, 2(3); 164–167: 2011.

❏❏❏

2

Classification of Biodegradable Polymers

> **INSIDE THIS CHAPTER**
>
> 2.0 Categories of Polymers
>
> 2.1 Non-degradable and Non-renewable Polymers
>
> 2.2 Degradable and Non-renewable Polymers
>
> 2.3 Biodegradable and Non-renewable Polymers Drawbacks of Biodegradable and Renewable Polymers
>
> 2.4 Biodegradable and Renewable Polymers Drawbacks of Biodegradable and Renewable Polymers
>
> 2.5 Categories of Biodegradable Polymers
>
> 2.6 Details of Different Categories Biodegradable Polymers
>
> 2.7 References

Development of Biodegradable polymers has become an interesting field due to environmental concerns of disposal of synthetic polymers in landfill sites requiring precious space on earth and due to depleting petroleum resources required to fulfil the day to day requirements of human being. To understand the properties and application of biodegradable polymers which are found in nature or are being synthesised by the conventional methods, there is a need to understand the entire scenario of biodegradable polymers developed till date using various methodologies. In order to do so, first we need to understand the categories of polymers which are described here. Thereafter various classifications of different biodegradable polymers proposed by researchers have been discussed in detail.

2.0 CATEGORIES OF POLYMERS

In today's scenario polymers can be broadly classified in four categories. This classification is based on type of resources available in our universe.

- Non degradable and non-renewable
    - Traditional Petrochemical based plastics
- Degradable and non-renewable
    - From Petrochemical resources
    - Additives to break down Polymer chains
- Biodegradable and non-renewable
    - From Petrochemical resources

– Biodegradable and Compostable
- Biodegradable and Renewable
   – From Renewable Resources
   – 100% Biodegradable and Compostable

2.1 NON-DEGRADABLE AND NON-RENEWABLE POLYMERS

These polymers are derived from non renewable resources. Traditional petro chemicals based plastics/polymers fall in this category. Engineering plastics and commodity plastics are well known examples of this category. Engineering plastics are the plastics material which have superior thermal and mechanical properties and are costlier while commodity plastics are the group of plastics material which exhibit high production volume products. Category of petrochemical based non degradable polymers are:

1. Engineering plastics
2. Commodity plastics

Examples of Engineering plastics are:
- Polycarbonates (PC)
- Polyamides (PA)
- Polybutylene terephthalate (PBT)
- Polyethylene terephthalate (PET)
- Polyphenylene oxide (PPO)
- Acrylonitrile butadiene styrene (ABS) etc.

Examples of Commodity plastics are:
- Polystyrene (PS),
- Polyvinyl chloride (PVC),
- Polypropylene (PP), and
- Polyethylene (PE) etc.

Drawbacks of Non-Degradable and Non-Renewable Polymers

These are the polymers which have been derived from non renewable resources. These non renewable resources are undergoing fast depletion and as per the recent estimated the stocks will last not more than 70 years. This is a fact that these polymers do not biodegrade and remain in the environment for a very long time.

2.2 DEGRADABLE AND NON-RENEWABLE POLYMERS

These polymers are also derived from non-renewable resources. Some **additives are used to break down Polymer chains.** The use of additives has been done to make these polymers degradable but under certain controlled environmental conditions. Additives impart controlled degradation behaviour in conventional thermoplastics and known as pro-degradant concentrates. These are generally catalytic transition metal compounds. Examples are: Cobalt Stearate and Manganese Stearate. Generally an amount of 1–3% of additives is used, due to which cost of polymer gets increased in between 10 to 35%. Polyethylene containing 3% of the additive is claimed to degrade to 95% weight loss after four (4) weeks at 60°C. Hence, in order to breakdown this material the waste has to be heated upto 60°C continuously for very long time of the order of four to eight weeks. Thereafter the material will decompose. This is not practically possible as it would require to collect waste syregate and heat up through application of external energy to be supplied and process will become costly.

Drawbacks of Degradable and Non-Renewable Polymers

1. It has to be a Controlled degradation in a certain environmental conditions.

2. Catalytic metals used as additives remains in the environment after polymer degradation. Which affect adversely to our environment.

3. These non renewable polymers disintegrate into tiny particles and do not truly biodegrade. The small fragments remaining in the environment have too high a molecular weight to break down to low molecular weight materials.

2.3 BIODEGRADABLE AND NON-RENEWABLE POLYMERS

- These are derived from Petrochemical resources
- These are Biodegradable and Compostable.

These polymers are prepared from petrochemical resources by conventional synthesis method like polymerization from non-renewable monomer feedstocks. These are synthetic polymers, however, biodegradable. Such types of polymers are very expensive and generally used for medical applications.

Categories of synthetic polymers are:

1. Polyesters

    These are generally prepared from diols (Ethylene glycol, 1, 4-butanediols, 1, 4 hexandiol, 1, 4 Cyclohexane dimethanol),

2.3 Biodegradable and non-renewable Polymers

dicarboxlic acids (succinic acid, sabacic acid, terepthalic acid etc.). These polyesters can be further classified as:

- Aliphatic polyesters & Co polyester:
- PBS-Poly (butylene succinate),
- PBSA-Poly (butylene succinate adipate)
- Aromatic Co polyester:
- PBAT-Poly (butylene adipate terephthalate).

Examples:

- BASF (Germany) introduced **Ecoflex** Copolymer which contains adipic acid, terephthalic acid and 1,4 Butanediol.
- Mitsui Toatsu (Japan) introduced poly cyclohexane dimethanoyl succinate copolymer which contains diol and diacid.
- Eastman Chemical (USA) developed **Eastar** which contains 1,4 Butane diol, adipic acid, and terephthalic acid.

2. Polyesteramide

Example:

- BAYER introduced biodegradable polyesteramide as a thermoplastic with melting point 125°C.

3. Poly (ε-Caprolactone)

Example:

- Union Carbide introduced **Tone Polymer** which contained polycaprolactone. It is semi-crystalline thermoplastic linear aliphatic polyester.

4. Polyvinyl alcohol (PVA)

– It is semi crystalline carbon backbone polymer available in two forms.

(*i*) Commercial grade water soluble PVA containing degree of hydrolysis 88%.

(*ii*) Commercial grade water insoluble PVA containing degree of hydrolysis more than 98%.

Drawbacks of Biodegradable and Non-Renewable Polymers

These polymers are readily degradable by micro-organism but are very expensive and can not be used for short term daily use applications. Hence, their use is limited to medical applications only.

2.4 BIODEGRADABLE AND RENEWABLE POLYMERS

- Derived from Renewable Resources
- 100% Biodegradable and Compostable

These polymers are prepared from renewable resources which are produced in nature by living organism as well as from agricultural resources. These are based on truly renewable resources and considered as environmentally acceptable biodegradable polymers. The most wide spread biodegradable polymers categories are:

1. Polysaccharadies

   - Starch
   - Cellulose
   - Chitin and Chitosan

2. Protein
    - Corn Zein Protein
    - Soy Protein
    - Milk Protein
    - Collagen and Gelatin
    - Wheat Gluten Protein
3. Bacterial Polyester
    - Poly hydroxy alkanoates
4. Lipid
5. Vegetable oil
6. Hemi-cellulose
7. Poly lactic acid (PLA)

Drawbacks of Biodegradable and Renewable Polymers

These polymers are 100% biodegradable and readily available in our environment. Some of them are very cheap but they do not have desired properties like synthetic polymers. At the same time processing of these polymers is not easy to incorporate desired properties. Some of the polymers are derived from agro based biodegradable materials in which monomer is derived from the fermentation process but polymerization is very expensive.

2.5 CATEGORIES OF BIODEGRADABLE POLYMERS

The classification of Biodegradable polymers have also been done by various researchers based on various factors involved like material structure, origin, nature, processing methods etc.

Accordingly classifications have been summarized here.

1. Biodegradable polymers based on resources

    (*a*) Renewable resources based

    (*i*) Vegetable resources

    1. Wood

        (*a*) Cellulose

        (*b*) Lignin

        (*c*) Hemi cellulose

        (*d*) Natural rubber

        (*e*) Wood resin

        (*f*) Terpene

    2. Plants

        (*a*) Starch

        (*b*) Vegetable oil

        (*c*) Hemi cellulose

    (*ii*) Animal resources

    1. Chitin and Chitosan

    2. Protein

    3. Cellulose base fiber

    (*iii*) Bacterial resources

    1. Poly hydroxyl alkanoates

    2. Poly lactic acid

    3. Bacterial cellulose

2.5 Categories of Biodegradable Polymers

  (*b*) Non renewable resources based

   (*i*) Poly aliphatic ester

   (*ii*) Poly aromatic ester

   (*iii*) Poly esteramide

   (*iv*) Poly (Caprolactone)

  (*c*) Mixed resources based

   (*i*) Starch/Poly (Vinyl Alcohol) (PVA) Based Biodegradable Polymers

   (*ii*) Starch/Poly (ε-Caprolactone) (PCL) Based Biodegradable Polymers

   (*iii*) Starch/Poly (D, L-Lactic Acid) (PLA) Based Biodegradable Polymers

   (*iv*) Starch/Poly-hydroxy-alkanoates (PHA) Based Biodegradable Polymers

   (*v*) Starch/Poly (Butylenes Succinate-Co-Butylene Adipate) (PBSA) Based Biodegradable Polymers

2. Biodegradable polymers based on origin

  (*a*) Natural-Starch, Cellulose, Protein, PHA

  (*b*) Synthetic – PCL, PVA, PBSA, PLA

3. Biodegradable polymers based on polymeric Nature

  (*a*) Thermoplastic – Starch, Cellulose, PLA

  (*b*) Thermosets – Triglycerides, Epoxy resins and curing agents.

2.6 DETAILS OF DIFFERENT CATEGORIES BASED BIODEGRADABLE POLYMERS

As detailed above, there are different permutation and combinations based on which biodegradable polymers have been categorized.

1. **Biodegradable polymers based on resources:** This classification is based on available resources which are renewable and non-renewable.

    (*a*) **Renewable resources based:** Renewable resources based biodegradable polymer includes polysaccharides (such as cellulose, starch), protein (*e.g.* silk), as well as Polyhydroxyalkanoates which are synthesized by bacteria. They are generally produced in nature by living organisms and are a truly renewable resource. They are considered as environmentally acceptable degradable polymers. Renewable resources based biodegradable polymers are broadly classified into three categories *i.e.* vegetable, animal and bacterial resources based Biodegradable polymers

    (*i*) **Vegetable resources:** As per an estimation world biomass of vegetable is nearly 10^{13} tons and that solar energy renewes about 3% of it per annum[1]. The fundamental role of biomass is maintenance of oxygen level, principle of sustainiability limits its exploitation at most to that renewed percentage. Vegetable resources play a significant role for the development of Biodegradable polymers. These are readily available at low cost and available in abundance. Main category of vegetable resources based biodegradable polymers are wood, plants.

2.6 Details of different categories Based Biodegradable Polymers

1. Wood

    (*a*) Cellulose

    (*b*) Lignin

    (*c*) Hemi cellulose

    (*d*) Natural rubber

    (*e*) Wood resin

    (*f*) Terpene

2. Plants

    (*a*) Starch

    (*b*) Vegetable oil

    (*c*) Hemi cellulose

(*ii*) **Animal resources:** Biodegradable polymers can be extracted as polysaccharides from the exo skeleton of insects and outer skin of fungi, as protein from spider, dragline silk (with extra ordinary mechanical properties) as whiskers and Nanofibrils (with regular nano rods having extra ordinary mechanical and rheo-modifying properties) from tunicate mollusk. Main biodegradable polymers produced from animal resources are:

1. Chitin and Chitosan

2. Protein

3. Cellulose base fiber

(*iii*) **Bacterial resources:** These are naturally occurring biodegradable polymers. These polymers are susceptible to microbial degradation; the micro-organism secretes enzymes that attack the polymer and cleave it into smaller segments amenable to metabolization by microbial flora. Biodegradable polymers can be obtained by microbial production (PHA) and chemically synthesized through monomers obtained from agro based resources using micro-organism through fermentation process (PLA). Bacterial cellulose is obtained through biotechnological production having fibrous morphology. It is a promising field for further development of biodegradable polymers. Main biodegradable polymers produced from bacterial resources are:

1. Poly hydroxyl alkanoates (PHA)

2. Poly lactic acid (PLA)

3. Bacterial cellulose.

(*b*) Non renewable resources based: Synthetic biodegradable polymers are only produced by mankind. The major category consists of polyester with hydrolysable linkage along the polymer chain backbone. Biodegradable polymers are prepared from petroleum based material, whose monomers and polymers are obtained conventionally by chemical

2.6 Details of different categories Based Biodegradable Polymers

synthesis. These are generally prepared by poly-condensation, ring opening polymerization, and enzymes catalyzed polymer synthesis method using petrochemical feedstock as raw material. Traditionally polyester is prepared through poly-condensation of diols and diacids or from hydroxyl acids. Main biodegradable polymers produced from non renewable resources are:

(*i*) Poly aliphatic ester

(*ii*) Polyaromatic ester

(*iii*) Polyesteramide (PEA)

(*iv*) Poly (Caprolactone) (PCL)

(*v*) Poly Vinyl alcohol (PVOH).

(*c*) **Mixed resources based polymers:** One of the strategy adopted in preparing biodegradable polymers is compounding and blending of two different kinds of polymers. In day to day life most people are interested in using green materials but do not want to spend more money or use materials having inferior performance than the existing dominant fossil fuel based polymers and materials. Currently it is difficult to replace petroleum-based materials, from a cost and performance perspective. It is not necessary to make a 100% substitution for petroleum based materials immediately. A viable solution is to combine the different features and benefits of both petroleum and bio-resources to produce a useful product having the requisite cost-performance properties for real-world applications. Mixed resource based biodegradable polymers may be prepared by the compounding of:

1. Two/more biodegradable polymers (*e.g.* Starch + PLA)

2. One natural polymer + one petrochemical based polymer (*e.g.* Starch + Plastic)

Main biodegradable polymers produced from mixed resources are:

(*i*) Starch/Poly (Vinyl Alcohol) (PVA) Based Biodegradable Polymers

(*ii*) Starch/Poly (ε-Caprolactone) (PCL) Based Biodegradable Polymers

(*iii*) Starch/Poly (D, L-Lactic Acid) (PLA) Based Biodegradable Polymers

(*iv*) Starch/Poly-hydroxy-alkanoates (PHA) Based Biodegradable Polymers

(*v*) Starch/Poly (Butylenes Succinate-Co-Butylene Adipate) (PBSA) Based Biodegradable Polymers.

2. **Biodegradable polymers based on origin**

   (*a*) **Natural Biodegradable Polymers:** Natural Biodegradable polymers are readily available in nature. They are found in plants, animals and as well as in lower organism like bacteria. Natural polymers

2.6 Details of different categories Based Biodegradable Polymers

are produced in the growth cycles of cells of living organisms. Their synthesis includes enzymes catalysed polymerisation reaction of activated monomers, which occurs within cells as product of metabolic process. The most beautiful aspect of natural polymer is that they get degraded naturally. All natural polymers have energy and matter stored in them which is in decomposition released and made available for reuse at any point of time. These are considered as environmentally acceptable degradable polymers. Main Natural Biodegradable polymer are:

(i) Polysaccharides: This family is represented by different products such as starch and cellulose based on glucose units linked in macro molecular chain. Polysaccharides are high molecular mass carbohydrates. The molecular mass range of polysaccharides is wide. Polysaccharides are generally based on agricultural products and biological products.

Principal polysaccharides of interest are starch, and cellulose but other biopolymers are also been considered as biopolymers belonging to this category such as chitin and chitosen. The physical properties of polysaccharides vary widely. The globular and lower molecular mass polysaccharides (*i.e.* inulin or amylose) are water soluble, however, linear fibrous high molecular mass polysaccharides only swell in water and at the most form colloidal, often highly viscous or gelling, solutions.

Water soluble polysaccharides swell in water, giving colloidal, highly viscous solutions or pseudoplastic flow properties. Functional properties such as thickening, water holding and binding, stabilization of suspensions and emulsions, and gelling are based on this behaviour. Therefore, polysaccharides are often referred to as gelling or thickening agents, stabilizers, water binders, or fillers. Main examples of polysaccharides are:

1. Cellulose

2. Starch

3. Chitin and Chitosen

(*ii*) **Proteins:** A number of raw materials from agricultural resources have been used to produce renewable, biodegradable, and edible packaging polymeric materials. Proteins have been defined as natural polymer able to form amorphous three dimensional structures stabilized mainly by non covalent interactions. The functional properties of these materials are highly dependent on structural heterogeneity, thermal sensitivity, and hydrophilic behaviour of proteins. Main example of Protein are

1. Proteins from plant origin:

   (*a*) Gluten proteins

   (*b*) Soy proteins

   (*c*) Corn zein proteins

2.6 Details of different categories Based Biodegradable Polymers

    (*d*) Wheat gluten proteins

    (*e*) Peanut proteins

 2. Proteins from animal origin are:

    (*a*) Silk proteins

    (*b*) Milk proteins

    (*c*) Collagen (Gelatin) proteins

    (*d*) Casein

    (*e*) Keratin.

(*iii*) **Polyesters:** Polyesters are synthesized by bacteria. They are generally produced in nature by living organisms and are truly renewable resource. PHA's are linear, homochiral, thermoplastic polyesters produced as intracellular energy reserves–in effect, fat deposits-numerous micro-organisms. These polymers accumulate as distinct granular inclusions in response to nutrient limitations. The micro-organisms can also enzymatically degrade these granules when limitation is removed. The nature has provided us natural, renewable, biodegradable polyesters. Examples are:

1. Polyhydroxybutyrate (PHA)

2. Polyhydroxyvalerate (PHV).

(*b*) **Synthetic Biodegradable Polymers**: While natural polymers are produced by living organism, synthetic biodegradable polymers are only produced by mankind. The major category consists of polyester with hydrolysable linkage along the polymer chain backbone. Main synthetic biodegradable polymers are:

(1) Poly (Lactic acid)

(2) Poly (Glycolic acid)

(3) Poly caprocatone

(4) Poly (Vinyl alcohol)

3. **Biodegradable polymers based on thermal Behaviour**: Many different polymers have been synthesised and more will be produced in future. All bio-degradable polymers can also be assigned to one of two groups similar to classification of polymers based on processing characteristics or the type of polymerization mechanism. Based on thermal processing behaviour these are divided into two categories. These polymers that can be heat softened in order to process into a derived form called thermoplastics. Thermosets are polymers whose individual chains have been chemically linked by covalant bonds during polymerization or by subsequent chemical or thermal treatment during fabrications. Once formed they resist heat softening, mechanical deformation and solvent attack but can not be thermally processed.

  (*a*) *Thermoplastic:* Starch, Cellulose, PLA, PVA, PCL, PHA.

  (*b*) *Thermosets:* Triglycerides, Epoxy resins and curing agents.

2.7 REFERENCES

[1] Handbook of biopolymers and biodegradable plastics, properties, processing and applications, Sina Ebnesajjad, Chapter No. 4 A. Gandini and M.N. Belgacem. The State of art polymer from renewable resources, Elsevier, p-73, 2013.

[2] Polymer Science and Technology, 3rd edition, Joel R Fried, 2014, Prentice Hall Properties, Synthesis, Application and simulations.

❏❏❏

3

Materials and Their Structures

INSIDE THIS CHAPTER

3.1 Polysaccharides

    3.1.1 Starch

        3.1.1.1 Composition of Starch

        3.1.1.2 Properties of Starch

        3.1.1.3 Modifications of Starch

    3.1.2 Cellulose

    3.1.3 Chitin and Chitosen

3.2 Protein- Based

3.3 Bacterial/Microbial Polyesters

3.4 Poly (Glycolic Acid)

3.5 Synthetic Polyester

    3.5.1 Polycaprolactone

    3.5.2 Poly(Lactic Acid)

3.6 Poly(Vinyl Alcohol)

Contd...

3.7 Starch/Poly(Vinyl Alcohol) (PVA) Based Biodegradable Polymers

3.8 Starch/Poly(Epsilon-Caprolactone) (PCL) Based Biodegradable Polymers

3.9 Starch/Poly(D, L- Lactic Acid) (PLA) Based Biodegradable Polymers

3.10 Starch/ Poly-hydroxy-alkanoates (PHA) Based Biodegradable Polymers

3.11 Starch/Poly (Butylenes Succinate-Co-Butylene Adipate) (PBSA) Based Biodegradable Polymers

3.12 Starch with Olefins Vinyl Derivatives

3.13 Starch and Olefins with Acrylate Derivatives

3.14 Starch & Olefins with Anhydride Group

3.15 References

3.0 BIODEGRADABLE POLYMER - MATERIALS AND THEIR STRUCTURES

3.1 POLYSACCHARIDES

Polysaccharides are high molecular mass carbohydrates and most often insoluble in water. Like oligosaccharides, they are formed by linking mono-saccharides together through glycosidic linkage. The molecular mass range of polysaccharides is wide, between approximately five thousand for inulin and several million for glycogen or gum Arabic. Polysaccharides are generally based on agricultural products (*i.e.* cellulose, starch, konjac, lignin etc.) and biological products (*i.e.* chitin and chitosan, pullulan etc.

The physical properties of polysaccharides vary widely. The globular and lower molecular mass polysaccharides (*i.e.* inulin or amylose) are water soluble, however linear fibrous high molecular mass polysaccharides only swell in water and at the most form colloidal, often highly viscous or gelling, solutions.

Lower molecular mars polysaccharides swell in water, giving colloidal, highly viscous solutions or pseudoplastic flow properties. Functional properties such as thickening, water holding and binding, stabilization of suspensions and emulsions, and gelling are based on this behaviour. Therefore, polysaccharides are often referred to as gelling or thickening agents, stabilizers, water binders, or fillers.

3.1.1 Starch

Starch, the principal source of dietary calories to the world's human population, has many chemical and physical characteristics that set it apart from other food components and give it numerous applications. Starch, the storage polysaccharides of cereals, legumes and tubers, is a renewable and widely available raw material which can be used for various industrial applications. Starch is utilised in paper manufacturing as a wet-end retention aid as surface sizing and paper coating. Starch converted to thermoplastic material offers an alternative for synthetic polymers used in various applications. To make use of its properties, it is important to know its structural and behavioral characteristics. Unlike other carbohydrates and edible polymers, starch occurs as discrete particle in granules form. These particles are unique to the plant source and, therefore, starch from every plant type is different in appearance, properties and particle size distribution as shown in Table 3.1 [1]. When biodegradability is required, the potential applications are in non food areas, like capsule, agricultural films, packaging films for short term use, seed coatings, disposal bags, etc. These products are already available in the market. Starch is a potentially useful polymer for thermoplastic biodegradable materials because of its low cost, availability, and production from annually renewable resources. Starch has processing difficulties because of its T_g and T_m above decomposition temperature. This does not allow processing easily and requires processing before hand. However this difficulty could be overcome by using plasticiser such as water which is when added to starch will lower its T_g and T_m below the decomposition temperature.

3.1.1.1 Composition of Starch

Starch in granule form is generally composed of two types of molecules amylose and amylopectin.

(*a*) Amylose is a linear or sparsely branched carbohydrate based on α (1–4) bonds with a molecular weight of 10^5 to 10^6. The chains show spiral shaped single or double helixes.

Table 3.1: Compositions and Characteristics of Different Kinds of Starch

Source	Amylose Contents (%)	Amylopectin Contents (%)	Phosphorous Contents (%)	Moisture Contents (%)	Mean diameter, μm	Crystallinity (%)
Wheat	26-27	70-73	0.06	13	25	36
Maize	26-28%	71-73	0.01	12-13	15	39
Waxy Starch	<1	99	0.01	ND	15	39
Amylo-maize	50-80	22-50	0.03	ND	10	19
Potato	20-25	79-74	0.08	18-19	40-100	25

(*b*) Amylopectin is a highly multiple branched polymer with a high molecular weight 10^7 to 10^9. It is based on α (1–4) bonds but also on α (1–6) links constituting branching points occurring every 20 to 70 glucose units. The structures of amylose and amylopectin and representative structures are shown in Fig. 3.1, Fig. 3.2, Fig. 3.3. Linear amylose molecules tend to form helices which are often double. Although the structure of amylopectin is thought to be branched and perhaps bushlike as shown in Fig. 3.4 [1–4] the molecules have what may be called as a tassel-on-a-string structure.

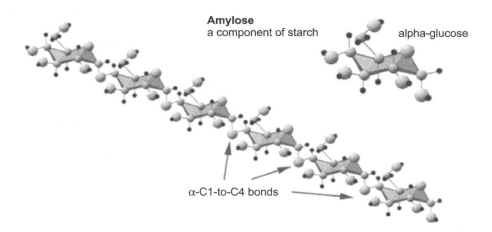

Fig. 3.1 Structure of Amylose

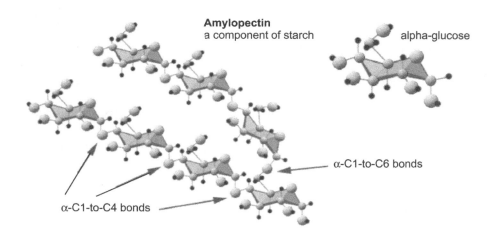

Fig. 3.2 Structure of Amylpectin

3.1 Polysaccharides

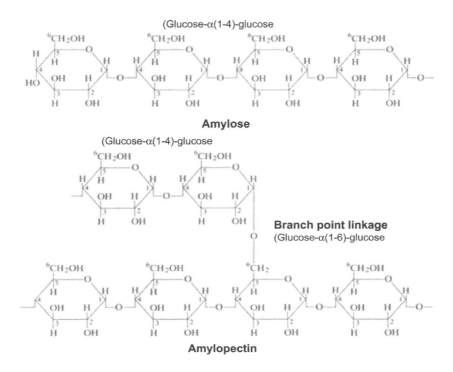

Fig. 3.3 Representative Structure of Starch

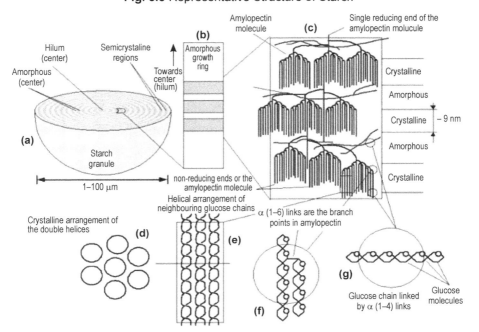

Fig. 3.4 Radial Structure of Starch Granules

3.1.1.2 Properties of Starch

Physicochemical properties of amylose and amylopectin are given in Table 3.2.

Table 3.2: Physicochemical Properties of Amylose and Amylopectin

Property	Amylose	Amylopectin
Molecular mass	5000-200000	one or several billion
Glycosidic linkages	Mainly (1,4)-α-D	mainly (1,4)-α-D, (1,6)-α-D
Susceptibility to retro-gradation	High	Low
Products of action of β-amylase	Maltose	maltose, β-limit dextrin
Products of action of glucoamylase	D-glucose	D-glucose
Molecular shape	essentially linear	bush shaped

3.1.1.3 Modifcations of Starch

Starch may be modified in numerous ways, both physically and chemically. The starch molecules, both in free and granular form, are subjected to chemical modification. Thus starch is modified in various ways to produce acid modified, oxidized, cross linked, partially esterified, or partially etherified starch or converted to cationic derivatives. These modifications, normally present at very low concentrations in starch preparations, produce dramatic differences in the physical and chemical properties.

(a) *Acid Modifed starch*: Acid treatment causes hydrolysis of glycosidic bonds in the starch modules. These are used in textile manufacturing.

(b) *Oxidized Starch*:

(i) Oxidation of reducing end aldehyde to carboxyl groups. Producing aldonic acid end units; namely, D-gluconic acid end units.

3.1 Polysaccharides

(ii) Oxidation of the C-6 Methynol group to a carboxyl group.

(iii) Oxidation of starch secondary hydroxyl to ketone group.

(iv) Oxidation of 2, 3-Glycol units to dialdehyde and dicarboxylic acid units (Glycol cleavage): Oxidative attack varies with the location of hydroxyl groups and with the nature of the oxidant. Oxidation at C-1, C-2, C-3 and C-6 determines the properties of the resulting starch product. Cleavage is shown in Fig. 3.5.

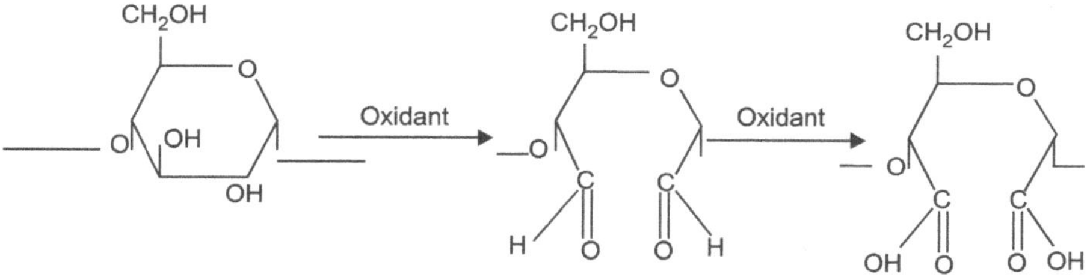

Fig. 3.5 Oxidation of 2,3-Glycol Units to Dialdehyde and Dicarboxylic Acid

(c) *Cross-linked Starch*: Cross linking of starch reinforces the intermolecular binding by introducing covalent bonds to supplement natural intermolecular hydrogen bonds. Cross linking restricts granule swelling, decreased peak viscosity on heating and increases the stability of gelatinized granules.

(d) *Starch Esters*: Carboxylate Esters and Phosphate Esters

(e) *Starch Ethers* : The two major types of commercially important starch ethers are hydroxy-ethyl and hydroxy-propyl ethers.

(f) *Cationic Starch*: Cationic starches include tertiary aminoalkyl ethers, quaternary ammonium ethers, amino-ethylated starches, cyanamide derivatives, starch anthranilates, and cationic dialdehyde starch.

(g) *Chemical Treatment*: Starch shows difficulties during thermoplastic processing. The starch has high melting temperature then its degradation temperature. Hence, during thermoplastic processing starch degrades instead of melting. To overcome this difficulty the starch has to be modified with chemicals like polyol, urea, sorbitol, amide etc. In due course of reactivity the melting temperature becomes low and the processing may be continued.

Examples of modified starch are as below:
- **Acetylated Starch**

  Starch acetates differing in the degree of substitution (from 1.5 to 2.5) were obtained from high-amylose corn starch. Mechanical properties of films prepared from acetylated starch are determined by the nature of native starch, being dependent on the content of amylose and amylopectin in the original polymer [5, 6]. These acetate films were less hygroscopic than starch films. These modified starches were less biodegradable, as esterification prevents the action of enzymes onto the starch by changing the nature of substrate.

- **Carboxymethyl Starch**

  Carboxymethyl starch had been obtained by the chemical treatment of yellow corn starch and pigweed starch (the degree of substitution = 0.1 – 0.2) with chloro-acetic acid [7, 8]. Generally, small amount of carboxymethyl starch used for the extrusion of various food stuff. It was found that modified starch improves processing of these products.

- **Hydroxy-propyl starch**

  Hydroxy-propyl starch mixed with gelatin and plasticized with polyols and water (up to 25 %) can be used to prepare elastic edible films.

- **Higher fatty acid esters of starch**

  Higher fatty acid esters of starch are promising reagents for mixing with non-polar polymers, such as PE and PP [9]. The reaction of starch with the corresponding acid chlorides in DMSO gave starch octanoates and dodecanoates with the DS of 1.8 and 2.7. The ester groups containing higher alkyl radicals improve the affinity of starch for the non-polar synthetic polymer and act as internal plasticizers. The compatibility of starch dodecanoate with PE is higher than that of starch octanoate. The starch dodecanoate-PE mixtures manifest higher thermal stabilities, greater elongation at break upon rupture and lower moisture absorption than starch octanoate-PE mixtures. However, being introduced into PE at any ratios, these substances decrease the rate of biodegradation of the mixture in comparison with starch-PE mixtures.

- **Starch modified by cholesterol**

  Starches modified by the introduction of cholesterol residues[10] and mixed with high-pressure PE were used for the production of blown films. In comparison with non-modified starch, these materials are more homogenous and more stable. Biodegradation in a compost of films prepared from these mixtures occurs even faster than that of non-modified starch-PE mixtures, apparently due to the loosening of the starch structure caused by bulky cholesterol fragments.

3.1.2 Cellulose

Cellulose is polysaccharide having a molecular structure similar to starch, however the D-glucose units are linked by β-glucosidic bonds in the cellulose. Cellulose is the most abundant renewable biopolymer on earth. It is understood that nearly one third of renewable matter on earth is cellulose. All the forms of cellulose are in the form semi crystalline fibers, whose morphology and aspect ratio can differ greatly from species to species. Cellulose is fully biodegradable, water – insoluble renewable resource. Cellulose esters and ethers, having a degree of substitution (DS) between 1.7 and 3.0 can be easily obtained from native cellulose. These materials have mechanical properties comparable to those of poly(styrene). Cellulose has multiple hydroxyl group because of which these are able to form hydrogen bond with oxygen molecule of another chain and holds the chain firmly together side by side by which these show exceptionally high tensile strength.

Cellulose natural fibers like bamboo, jute, hemp, flax, abaca are being used from several years as reinforcing polymer material. Cellulose fiber however, has hydrophobic nature, limited processing temperature, high moisture absorption, non uniform dimension and properties. Many of these disadvantages can be improved by the appropriate fiber treatment and processing.

3.1 Polysaccharides

By the addition of suitable plasticizer, they can be melt processed using the same techniques adopted for commodity thermoplastics polymers. Accordingly, cellulose derivatives have been utilized in the fabrication of textile fibers, photographic films, adhesive tapes, and dialysis membranes. Cellulose structure is shown in Fig. 3.6. Now a days cellulose fiber is used as the matrix to develop fully biodegradable composites (Green Composites).

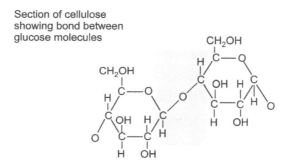

Fig. 3.6 Cellulose structure

- *Cellophane*: Basically cellophane is a regenerated cellulose produced by extruding a viscous alkaline colloidal dispersion of cellulose xanthate into an acid/salt bath. After treatment with plasticizer *i.e.* glycerol and after drying, a clear transparent film is obtained. Cellophane is brittle at dry conditions, hence a plasticizer was required to improve its flexibility. It possess higher vapour permeability but does not show thermoplastic property hence not heat sealable.

- *Cellulose Acetate* : Most abundant renewable natural polymer and excellent raw material for modifications to fulfill emerging needs. Cellulose acetate is another polysaccharide obtained by the reaction of cotton-cellulose with acetic anhydride. The production of cellulose ester from recycled paper and sugar cane has also been done and product had DS in between 1.7 to 3.0. Plasticizers were used to lower its processing temperature. CA films have tensile strength similar to polystyrene and hence can be injection molded. It is being used for clear adhesive tape, eyeglass frames, toothbrushes etc.

- *Hydroxy-propyl-methyl Cellulose*: It is a modified cellulose with many applications due to its varied and controllable properties like substitution and molecular weight control. It is used as a hydrophillic or water insoluble polymer in drug delivery system. It is non-ionic polymer exhibiting pH independent drug release profile. HPMC is a water soluble fiber that does not ferment in the stomach.

3.1.3 Chitin and Chitosan

It is the most abundant animal polysaccharides material on earth. It is found in exo-skeleton of insects and outer skin of fungi. Chemical structure of chitin is similar to cellulose which is regular linear polymer having differing structure than cellulose because of the presence of N-methylamide moieties in place of hydroxyl groups at C2. Chitin has NH-CO bond which has strong tendency to form intermolecular hydrogen bond. That is why it is sparingly soluble in the polar solvents. Chitin also degrade chemically below its melting point which makes it difficult in processing.

Chitosen is deacetylated chitin which is swollen with water and dissolves in a water acetic medium. When amide functional group of chitin is converted into primary amino group then compound is called chitosen. Average molecular weight of chitin is 1.03 to 2.5×10^6 Da, but upon N-deacetylation, it reduced to 1.0 to 5×10^5. Chitosen is soluble in dilute acids such as acetic acid, formic acid etc. This polymer is produced commercially from the base catalysed deacetylation of shellfish waste. Chitin and chitosen exhibit good mechanical properties as well as low permeabilities. Chitosen has also been found to exist naturally, being synthesized by zygomycete fungi as part of their cell wall. Chitosen is completely biodegraded by chitosanases enzyme. It has many useful properties as hydrophilicity, biocompatibility, biodegradability and antibacterial characteristics. It can form transparent film and fibers. Film was described as flexible, tough, transparent and colourless with a tensile strength of 6210kPa. Chitosan has prospective applications in many fields such as biomedicined, waste water treatment functional membranes and flocculation. It has also been used in the purification of drinking water, cosmetic products etc. beside several uses in medical applications such as wound dressing, drug delivery, encapsulation etc. Chitin and chitosen are shown in the Fig. 3.7.

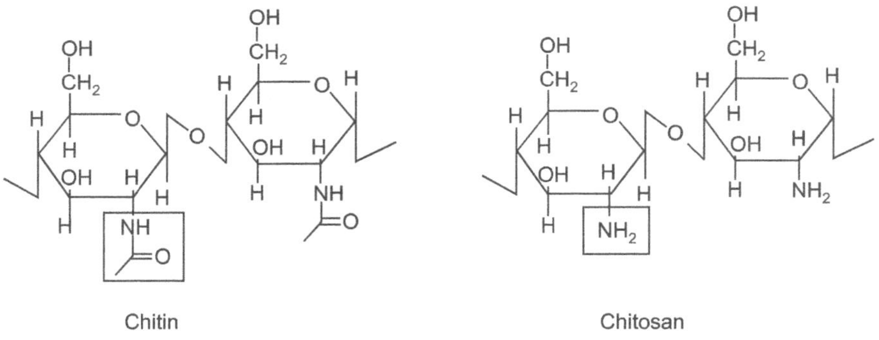

Fig. 3.7 Chemical structure of Chitin and Chitosen

3.2 PROTEIN- BASED

A number of raw materials from agricultural resources have been used to produce renewable, biodegradable, and edible packaging. Proteins have been defined as natural polymer able to form amorphous three dimensional structures stabilized mainly by non covalent interactions. The functional properties of proteins are dependent on structural heterogeneity, thermal sensitivity, and hydrophilic behaviour. Various vegetable and animal proteins, such as soy proteins, corn zein proteins, wheat gluten proteins, peanut proteins, milk proteins are used in packaging industry.

- **Soy Protein**

    Many researches have shown potential uses of soy plastics to replace petrochemical plastics for commodity applications such as food trays spoons, plastic bags [11–15]. These efforts have shown to be functionally successful, but they are still far from being considered, commercially attractive due to the higher raw material and processing costs.

    The molecular weight of soy protein is in between 20,000 to 35,000 dalton. Soy protein has an isoelectric point at about pH 4.5. At pH 4.5 the soy protein has the least net charge and thus is the most water resistant. It has been shown [14] that when pH drops from 6 to 4.5, water absorption of the plastics decreased from about 80% to 30% after 26h submersion in the water at 25°C. Soy protein has been chemically modified by reacting it with formaldehyde, glutaraldehyde, or acetic anhydride to make it better plastic [12]. The modern soy polymers are now in use as paper pigment-structuring agents and flow modifiers.

- **Corn Zein Proteins**

    Corn Zein Proteins are being used as a good renewable material for film forming. Injection-molded plastics display high water sensitivity, while cross-linked materials produce plastics with reduced water absorption and high mechanical properties [16].

- **Milk Proteins**

  Caseins and whey proteins are having film forming properties. Caseins dispersed in aqueous solution can form transparent, flexible, and tasteless films. Covalent cross-links catalyzed by transglutaminases could be formed to improve water resistance or to allow immobilization of active enzymes but these are not abundant.

- **Collagen and Gelatin**

  Collagen and Gelatin has been used for film forming. Collagen is used in the meat industry to form edible coatings through extrusion [17]. Gelatin has been used to form films which have properties like transparent, flexible, water resistant, and impermeability to oxygen. [18] but in comparison with starch it is costlier.

3.3 BACTERIAL/MICROBIAL POLYESTERS

These are naturally occurring biodegradable polymers which are susceptible to microbial degradation. The micro-organism secretes enzymes that attack the polymer and cleave it into smaller segments amenable to metabolization by microbial flora. Polyesters accumulated inside microbial cells as carbon and energy source storage.

- **Poly hydroxyalkanoates (PHA)**

  PHA's are linear, homochiral, thermoplastic polyesters produced as intracellular energy reserves – in effect, fat deposits-numerous micro-organisms. These polymers accumulate as distinct granular inclusions in response to nutrient limitations. The micro-organisms can also enzymatically degrade these granules on removal of applied limitations. The nature has provided us natural, renewable, biodegradable polyesters. Examples are Polyhydroxybutyrate (PHA) and Polyhydroxyvalerate (PHV).

3.4 POLY (GLYCOLIC ACID)

The first of the biodegradable synthetic polymer, poly(glycolic acid) (PGA; 2) that became commercially available [19] was chosen on the basis of results of a screening of potential materials subjected to degradation in vivo and in physiological solution over a period of 90 days [20]. The polymer was obtained by ring opening polymerization of glycolide, and the resultant polymer was extruded into a stiff fibre, with a high melting point of 225 °C, which degradaed in vitro in about 25 days [21]. Structure is given in Fig. 3.8.

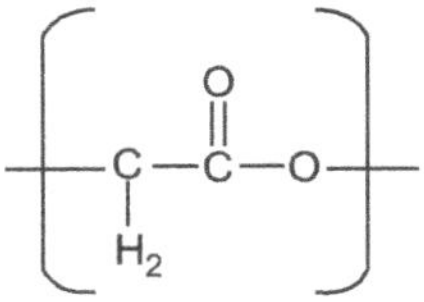

Fig. 3.8 Chemical Structure of Poly(Glycolic Acid)

3.5 SYNTHETIC POLYESTER

3.5.1 Polycaprolactone

PCL is an aliphatic polyester that is a well-known biodegradable polymer consists of hydrolyzable backbone of aliphatic polyester structure. The hydrolyzable backbone leads to good biodegradability and the aliphatic structure leads to good mechanical properties. Poly(caprolactone) (PCL) is obtained by the ring-opening polymerisation of ε-caprolactone. The structure of PCL is given in Fig. 3.9. The molecular weight of the polymer varies from 2000 to 80,000 daltons. PCL degrades quickly in soil burial, sludge and compost by micro-organisms. The low melting temperature of PCL is an advantage in accelerating degradation in composting environments; in which temperature often reaches 60°C. However, problems arise in applications in which high temperature is experienced that could compromise the mechanical integrity and its performance. Due to their high cost they are used in medical applications. Other uses of PCL include orthopedic casts, adhesive mold release agents, pigment dispersants, coatings and elastomers etc.

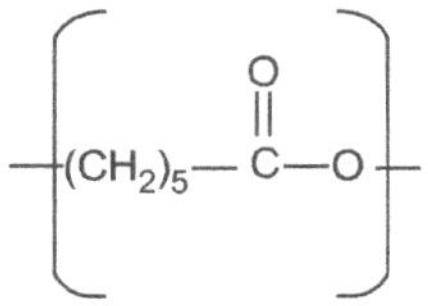

Fig. 3.9 Chemical Structure of Poly-caprolactone

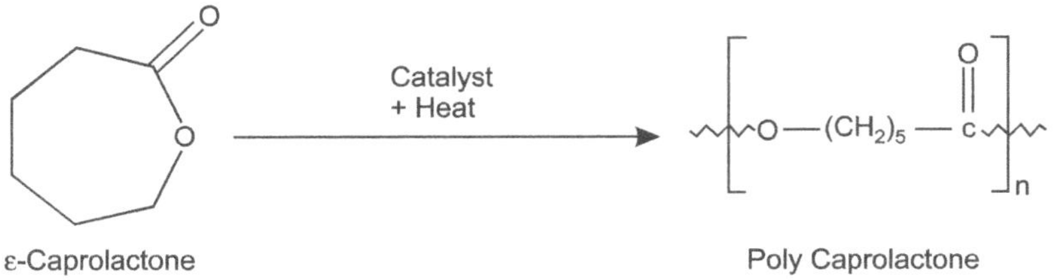

Fig. 3.10 Chemical reaction for preparation of Poly-caprolactone

3.5.2 Poly(Lactic Acid)

A lot of research and development activity is devoted to synthesize poly-lactic acid (PLA). Poly(lactic acid) or polylactide (PLA) is a thermoplastic aliphatic polyester commonly made from a-hydroxy acids, derived from renewable resources, such as

- corn starch (in the United States),
- roots, starch mostly in Asia,

   sugarcanes (in the rest of world).

It can biodegrade under certain conditions, such as the presence of oxygen, and is difficult to recycle. The structure of Poly(Lactic Acid) is given in Fig. 3.11 PLA is primarily used for medical applications, including drug delivery, vascular grafts, artificial skin and orthopedic implants, resorbable sutures, and prosthetic devices and as a vehicle for delivery of drugs and other bio active agents

It is a polyester, synthesized by the condensation polymerisation of lactides. Generally Bacterial fermentation is used to produce lactic acid from corn starch or cane sugar PLA of high molecular weight is produced from the dilactate ester by ring-opening polymerization Two lactic acid molecules undergo a single esterfication and then catalytically cyclized to make a cyclic lactide ester as shown in Fig. 3.12.

There are two isomers of PLA, the (L-)-lactide PLA and D, L-lactide. PLA produced commercially. The L form is highly crystalline, and the melt transit temperature decreases with lower molecular weight [22, 23]. The polymer of D, L-lactide is significantly more amorphous, with a correspondingly lower melting point. Polymerization of a racemic mixture of L- and D-lactides usually leads to the synthesis of poly-DL-lactide (PDLLA).

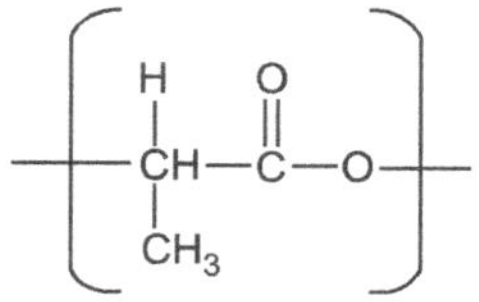

3.5 Synthetic Polyester

$$\left[\begin{array}{c} \underset{|}{H} \quad \underset{\|}{O} \\ -CH-C-O- \\ | \\ CH_3 \end{array}\right]$$

Fig. 3.11 Chemical Structure of Poly-lactic acid

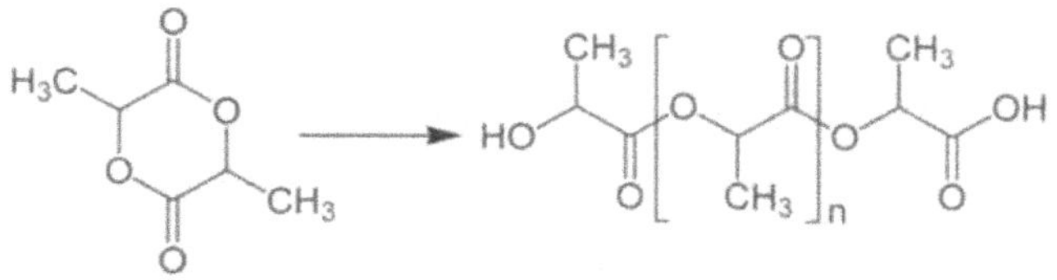

Fig. 3.12 Catalytic and thermolytic ring-opening polymerization of lactide (left) to polylactide (right)

PLA is considered both as biodegradable (*e.g.* adapted for short-term packaging) and as biocompatible in contact with living tissues (*e.g.* for biomedical applications such as implants, sutures, drug encapsulation, etc.).

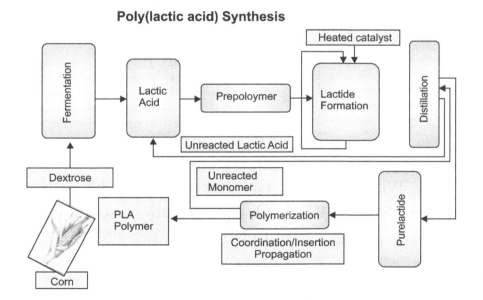

Fig. 3.13 Synthesis of PLA from Corn Starch

PLA can be degraded by abiotic degradation (*i.e.* simple hydrolysis of the ester bond without requiring the presence of enzymes to catalyze it). During biodegradation process, only in a second step, enzymes degrade the residual oligomers till final mineralization (biotic degradation).

As long as the basic monomers (lactic acid) are produced from renewable resources (carbohydrates) by fermentation, PLA complies with the rising worldwide concept of sustainable development and is classified as an environmentally friendly material.

3.6 POLY (VINYL ALCOHOL)

Poly (Vinyl Alcohol (PVOH) is shown in Fig. 3.14 is a readily biodegradable water-soluble polymer. PVOH, a hydrolyzed product of polyvinyl acetate, has excellent strength and flexibility. It has processing difficulties because the melting point of PVOH is around 230°C for the fully hydrolyzed grade and around 180°C – 190°C for the partially hydrolyzed grade.

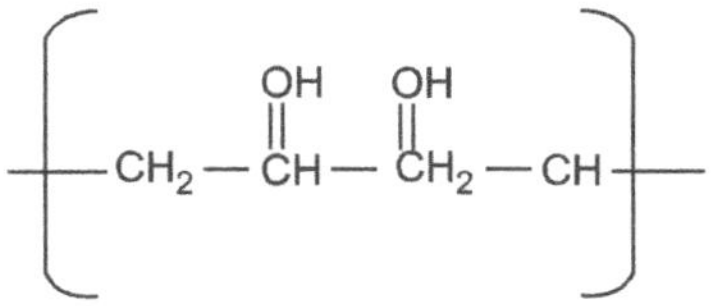

$$\left[-CH_2-\overset{OH}{\underset{\|}{CH}}-\overset{OH}{\underset{\|}{CH_2}}-CH- \right]$$

Fig. 3.14 Chemical Structure of Poly(Vinyl Alcohol)

The PVOH undergoes rapid decomposition above 200°C. Therefore, films PVOH can only be prepared by the solvent casting from water. Applications of PVOH include sizing binders, paper coatings, adhesives, fibers, and films for agricultural chemicals and hospital laundry bags. PVOH can be modified that retains its water solubility and it is extrudable through plastreizer. PVOH can be used in food processing industry. PVOH is bio-degradable involving two stage chain cleavage by secondary alcohol oxidase followed by beta-diketon hydrolase.

3.7 STARCH/POLY (VINYL ALCOHOL) (PVA) BASED BIODEGRADABLE POLYMERS

Starch-PVA films have been prepared by gelatinizing starch with PVA in the presence of plasticizers, and cross-linking agent, such as formaldehyde. It is hard to melt-process PVA because its thermo-degradation temperature is slightly higher than its melting temperature. Therefore, PVA has generally been used in solution forming, for example, in soluble gel and solution spinning. In general, the presence of PVA improves mechanical properties, and water resistance of the starch/PVA material, but humidity has more influence on these performances [24]. Relative absorption rates of starch-PVA films was similar to starch under higher relative humidity conditions while under low humidity conditions it showed an overall trend similar to that of PVA alone. At low humidity, absorption behavior of starch-PVA films is quite similar to pure PVA, which may be due to moisture barrier properties of PVA. The tensile strength of the starch PVA films decreases with the increase in relative humidity. The elongation of starch-PVA films can be improved by adding a small amount of poly (ethylene-co-acrylic acid) (EAA) [25]. A small amount (ca.3%) of EAA was optimum for obtaining films with 100% elongation and tensile strength of 25 MPa, while maintaining the starch contents above 50%. The solution processing cost and low processing efficiency are the main problems because of which the solution provided is not really acceptable. Thereafter it was [26] attempted to melt process through conventional processing techniques, with mixture of water and glycerol in 50:50 by weight, as optimum level of plasticizer. Here, mixing of PVA into starch could enhance the mechanical properties but the improvement is limited which may be attributed to the poor interface adhesion between the fibrous PVA structure and the starch matrix.

The Starch-PVA biodegradable polymer has been used as water-soluble laundry bags. Although the starch-PVA composition exhibits desirable properties in the beginning, their quality deteriorates fairly rapidly with time. When strict biodegradation and water resistance is required, this combination had no better properties.

3.8 STARCH/POLY (EPSILON-CAPROLACTONE) (PCL) BASED BIODEGRADABLE POLYMERS[66]

PCL show good mechanical properties, compatibility with many types of polymers commercially available. The incorporation of starch in it provides very useful viable products which would biodegrade, and would be of low cost since, the starch is available in abundance in our country. [27–33]. Generally starch and PCL do not exhibit significant interfacial adhesion, to improve the compatibility the modified polycaprolactone is being used. Modified polycaprolactone (PCL-g-methacrylate, GPCL) with gelatinized starch, polycaprolactone modified by pyromellitic anhydride and Starch, starch-g-PCL (SGCL) graft co-polymer as a compatibilizer have been used for improvement of the compatibility. These systems show good interfacial adhesion that exhibit better strength properties [31, 34–37].

3.9 STARCH/POLY(D, L- LACTIC ACID) (PLA) BASED BIODEGRADABLE POLYMERS

PLA is a brittle material with low possible elongation, which has a glass transition temperature of 54°C. The introduction of starch into such an already brittle matrix, results in even more brittle materials. There are few studies about PLA/Starch composites [38–41]. Introduction of low molecular weight poly(ethylene glycol) as a plasticizer into the PLA matrix can lower the glass transition temperature of PLA significantly and enhance the crystallization. When system was mixed as starch 40% and PEG as 20% then three component system is formed and showed further reduction in glass transition temperature and enhanced the crystallization.

Simple blending of starch and PLA does not impact upon T_g [42]. The tensile strength and elongation at break of the blends decreased as the starch contents increased. Addition of gelatinized starch, prepared with various contents of glycerol have decreased the crystallization temperature and the degree of crystallinity. Park *et al.* pointed out that the starch has acted as nucleating agent [43]. Gelatinization of starch has destroyed the hydrogen bonding in granules and decreased crystallinity. The introduction of glycerol during gelatinization prevented re-formation of hydrogen bonding between the starch polymeric chains. Gelatinized starch improved the interfacial adhesion between starch and PLA. For PLA/gelatinized starch blends, mechanical properties were superior to those for the PLA/pure starch blends.

3.10 STARCH / POLY-HYDROXY-ALKANOATES (PHA) BASED BIODEGRADABLE POLYMERS

Bacterial Poly(hydroxyalkanoates) (PHA's) shown in Fig. 3.16 are represented by poly (3-hydroxy butyrate-co-3-hydroxyvalerate)

(PHBV (C), in Fig. 3.17). PHBV has good mechanical qualities. Different test environments give different biodegradation results of PHBV. It undergoes quite rapid enzymatic hydrolysis in soil, sewage sludge and seawater, especially in the presence of extracellular P(3HB) depolymers, but the polymer is hydrophobic and anaerobic hydrolysis is relatively slow [44] as informed. Many factors such as surface area, temperature, microbial density and composition, enzymatic percolation, microbe infiltration, glucose repression of PHA esterase activity, etc., influence the degradation rate at which PHBV gets degraded [45]. Gorden *et al.* [46] has proposed the mathematical model to predict the bioplastics weight loss profile when starch and PHBV are mixed. The method of loose-filling of starch and cellulose acetate into PHAs matrix have been investigated by some researchers [47–50]. The degree of adhesion between starch granules and PHBV is poor, but can be mitigated via appropriate formulation and processing techniques, so as to obtain material with commercially useful properties. It is also reported that

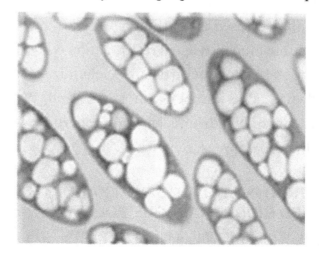

Fig. 3.15 Intracellular PHA storage

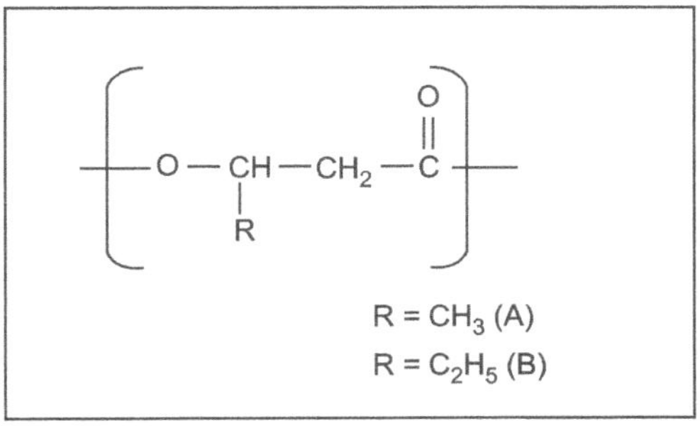

3.10 Starch/Poly-hydroxy-alkanoates (PHA) Based Biodegradable Polymers

Fig. 3.16 Chemical Structure of PHA's (PHB (A) & PHV(B))

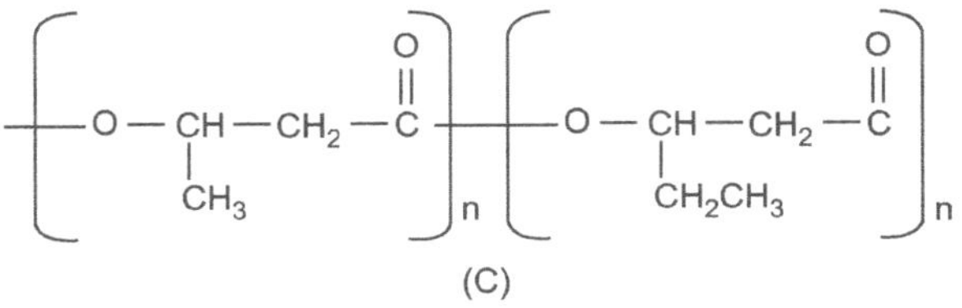

Fig. 3.17 Chemical Structure of PHBV (C)

pre-coating of starch with polyethylene oxide (PEO) had improved the tensile strength and elongation may be due to improved adhesion from PEO-PHBV interaction [51]. In compost such modification had no effect on biodegradation. [52]. However, in municipal sludge, incorporation of 9% PEO into starch-PHBV significantly slowed biodegradation [53] as reported. Willet *et al.* [50] have investigated the use of graft co-polymers of starch and glycidyl methacrylate (starch-g-PGMA) in composites that exhibit improved tensile and flexural strength compared to untreated starch controls. It has been reported that adhesion improved in the presence of starch-g-PGMA and also reported that even in the absence of chemical reaction, the presence of grafted PGMA on starch granules enhanced the adhesion with PHBV which improved mechanical properties, especially after absorption of water by starch granules.

3.11 STARCH/POLY(BUTYLENES SUCCINATE-CO-BUTYLENE ADIPATE) (PBSA) BASED BIODEGRADABLE POLYMERS

PBSA is commercially available aliphatic polyester, synthesized through glycol with aliphatic dicarboxylic acid. PBSA degrade slowly in the activated sludge soil. When starch is used as filler expectation was that the degradation rate would be improved thereby cost of PBSA would be reduced. In one of the studies it is reported that on mixing of 15, 30, 50% starch, the prepared films were completely consumed within 45, 30 and 20 days respectively, while 100% PBSA can not be consumed completely even after 60 days [54]. Ratto and co-workers [55] reported that PBSA and corn starch composites could be prepared up to 30% starch contents and films could be made. Although the addition of starch could alter the mechanical properties of the resulting film, these materials retained the toughness as well as the useful range of mechanical properties of the films. The PBSA is being used as trash bags. It was observed that moisture content did not affect the mechanical properties. The starch combination in PBSA results into highly improved biodegradation rate up to a certain range in the soil environment.

3.12 STARCH WITH OLEFINS VINYL DERIVATIVES

These type of biodegradable polymers are prepared from native corn as well as from high-amylose starch by extrusion with the

3.12 Starch with Olefins Vinyl Derivatives

Ethylene- Vinyl Alcohol Co-polymer [EVAL, 56% of $CH_2CH(OH)$ units [56].

Heterogeneous mixtures of EVAL with yellow corn starch displayed a clearly defined boundary between the phases; the sizes of domains observed were proportional to actual extrusion time. The phase separated mixtures of various compositions, containing domains (0.1–3 mm) involving either with starch or EVAL depending on the composition were prepared [57]. As on starch content of 55–60 %, the phase reversal was accompanied by drastic changes in the domain sizes. The water resistance and mechanical properties of starch mixtures are determined by the ratio of their constituents.

Absorption of moisture by films [58] prepared by blowing of mixtures of native corn starch pre-plasticized with glycerol and water and EVAL [1 : 1] (mixture A); [2 : 1] (mixture B) varied from 2% to 11%, was greater in the films prepared from mixture B. The breaking strength of the films prepared from mixture A was one-third of that for films prepared from pure EVAL and even less for films prepared from mixture B.

The conditions of processing starch mixtures with EVAL definitely influence degradability. The favorable effect of the structural anisotropy of starch – EVAL mixtures which appears in the injection moulding on the resistance of the mixtures to the action of physiological solutions and ethylene oxide used for the sterilization of materials in clinical practice was observed.

In-depth studies into the properties of starch mixtures with EVAC and with EVAC modified with maleic anhydride (EVAMA) were studied [6,7].

Mixtures of commercial corn starch (25% of amylose, 75% of amylopectin) with co-polymers containing 18% and 28% of CH_2-CHO-AC units and about 0.8% of anhydride fragments were prepared by extrusion at different temperatures, screw speeds and mixing times.

3.13 STARCH AND OLEFINS WITH ACRYLATE DERIVATIVES

Thermo-oxidation resistance of mixtures of LDPE with a Methyl Methacrylate – Butadiene-Styrene Co-polymer (MBS), with EAA and plasticized starch at 190°C was studied [83–85, 96–97] by using TGA, DSC and IR spectroscopy both upon continuous heating and in the isothermal regime. Conclusion was that MBS and EAA accelerated, whereas starch decelerated thermo-oxidation of LDPE. A ternary LDPE-EAA-starch system was found most resistant to temperature, due to the stabilizing effect of the co-polymer at the LDPE–starch interface.

3.14 STARCH AND OLEFINS WITH ANHYDRIDE GROUP

Mixtures of starch with an Ethylene–Propylene Co-polymer (EP), polystyrene (PS), an Ethylene-Propylene-Maleic Anhydride Co-polymer (0.8%, EPMA) as well as with a Styrene-Maleic Anhydride Co-polymer (8%, SMA) obtained by extrusion at 120–180°C, were studied at different effective mechanical moulding energies [233–235]. The effects of the composition (40-70% of starch) and moulding conditions on the structure and properties of starch mixtures were analysed. Miscibility of components in between Starch–EPMA mixtures, [59–64] has improved in the presence of anhydride fragments in the synthetic polymer chain. This phenomenon was interpreted [59, 61–62, 65] as the chemical reaction between starch and synthetic polymer under the influence of high temperatures and shear stress in extruder on moulding. This reaction results in the formation of ester bonds between the carboxy groups in the co-polymer chain and the primary hydroxy groups of starch. Starch mixtures with EPMA and SMA are easy to mould, display satisfactory mechanical characteristics and are biodegradable by spores of the fungus *Penicillium funiculogum*, which is facilitated with an increase in the starch content. At low starch content, starch granules remain encapsulated within the synthetic polymer and are thus hardly accessible to microorganisms.

3.15 REFERENCES

[1] Averous L, *J Macrom Sci Polym R C*, 44;231–274:2004

[2] Ray SS and Bousmina M, *Progr Mater Sci*, 50; 962–1079:2005.

[3] Rosa DS, Guedes CGF, Pedroso AG and Calil MR, *Mater Sci Eng C-BIO S C*, 24; 663–670:2004.

[4] Xiaofei Ma and Jiugao Yu, *Carbo Polym,* 57(2);197–203: 2004.

[5] H.J. Liu, L. Ramsden and H.Corke, *Carbo Polym*, 34; 283:1997.

[6] C. Fringant, J. Desbrieres and M.Rinaudo, *Polymer*, 37; 2663:1996.

[7] Mani, R., Bhattacharya, M., *Eur Polym J*, 37; 515–526:2000.

[8] M.Bhattacharya, R.S.Singhal and P.R. Kulkarni, *Carbo Polym*, 31;79:1996

[9] S. Thiebaud, J.Aburto, I.Alric, E. Borredon, D. Bikiaris, J. Prinos and C.Panayiotou, *J Appl Polym Sci*, 65; 705:1997.

[10] Bong Gyu Kang, Sung Hwa Yoon, Suck Hyun Lee, Jae Eue Yie, Byung Seon Yoon, Moon Ho Suh, *J Appl Polym Sci*, 60(11); 1977–1984:1998

[11] I. Pateau, C. Z. Chen, and J. Jane, *Ind Eng Chem Re* 33;1821:1994.

[12] I. Pateau, C. Z. Chen, and J. Jane, *J Env Polym Degrad*, 2; 211:1994.

[13] S. Wang, H. J. Sue, and J. Jane, *J Macro Sci Pure Appl Chem*, A33(5); 557:2006.

[14] N.C. Neilson, in New Protein Foods, Vol. 5 (A. M. Attschal and H.L. Wilke, Eds), Academic, New York, p-27:1985.

[15] H. J. Sue, S. Wang, and J. L. Jane, *Polymer*, 38;5035:1997.

[16] T. Herald, K.A. Hachmeister, S. Huang, and J.R. Bowers, *J Food Sci*, 61;415:1996.

[17] H..L. Wood, *Adv Meat Res*, 4;109:1987.

[18] G.D. Herbert, *U.S. Pat.* 3, 317, 631 1964

3.15 References

[19] E.E. Schmitt and R.A. Plistina, U.S. Pat. 3 297 033(1967) (Chem. Abstr., 1967, 66, 38656)

[20] E.J. Frazza and E.E. Schmitt, *Biomed Mater Symp*, 1971, I, 43.

[21] C. C. Chu, *Ann Surg*, 195; 55:1982.

[22] K. Jamshidi, S.H. Hyon, and Y. Ikada, *Polymer*, 29; 2229:1988.

[23] E.S. Lipinsky and R. G. Sinclair, *Chem Eng Prog*, 82(2); 26:1986.

[24] Chen. L., Imam, S. H., Stein, T. M., Gordon, S. H., Hou, C.T., Greene, R. V., *Polym Prepr*, 37;461–462:1996.

[25] Lawton, J. W. and Fanta, G. F., *Carbo Polym*, 23(4); 275–280:1994.

[26] Liu Z. Q., Feng Y., Yi X.S., *J Appl Polym Sci*, 74; 2667–2673:1999.

[27] Kweon D. K.; Lim, S. T., *J Appl Poly Sci*, 81; 2197-2202:2001.

[28] Oduasnya, O.S., Ishiaku U.S., Azsmi B. M. N., *Polym Eng Sci*, 40(6); 1298-1305:2000.

[29] Wang L., Shogren R. L., Carrier C., *Polym Eng Sci*, 40(2); 499–506:2000.

[30] Kim C. H., Cho. K.Y., Park J.K., *J Appl Polym Sci*, 81; 1507–1516:2001.

[31] Avella M., Errico M. E., Laurienzo P., Martudcelli E., Raimo M., Rimedio R., *Polymer*, 41; 3875–3881:2000.

[32] Vincent T. Breslin, Boen Li, *J Appl Polym Sci*, 48(12); 2063–2079:2003

[33] Jo Ann Ratto, Peter J. Stenhouse, Margaret Auerbach, John Mitchell and Richard Farrell, *Polymer*, 40(24); 6777–6788:1999

[34] Narayan R., Krishnan M., *ACS polym Mater Sci Eng* , 72; 186–187:1995.

[35] Kim C. H., Cho. K.Y., Park J.K., *J Appl Polym Sci* , 81; 507–1516: 2001.

[36] Choi E. J., Kim C.H., Park J. K., *J Poly Sci Part B: Polym Pysics* 37; 2430–2438:1999.

[37] Kim C.H., Cho K.Y., Park J. K., *Poly Eng Sci*, 41; 542–553: 2001.

[38] Fang, Q. and Hanna, M.A,. *Cereal Chem*, 77(6); 779–783: 2000.

[39] Jacobsen S., Fritz H. G., *Polym Eng Sci,* 36; 2799–2804: 1996.

[40] Park J. W., Lee D. J., Yoo E. S., Im S.S., Kim S. H., Kim Y. H. *Korea Polym J,* 7(2); 93-101: 1999.

[41] Martin, O., Averous, L., *polymer*, 42;6209-6219: 2001.

[42] KeT. Y. and Sun X. Z., *Cereal Chem*, 77; 232–236:2000.

[43] Park J.W. and Im S. S., *Polym Eng Sci*, 40; 2539–2550: 2000.

[44] Amass W., Amass A., Tighe B., *Polym I,* 47; 89–144: 1998.

[45] Imam S. H., Gordon S. H., Shogren R. L., Tosteson T. R., Govind N. S., Greene R. V., *App Environ Microbil*, 65(2); 431–437: 1999.

[46] Gordon S. H., Imam S.H., Shorgen R. L., Govind N. S., Greene R.V., *J Appl Polym Sci*, 76; 1767–1776:2000.

[47] Koenig M. F., Huang S.T., *Polymer*, 36; 1877–1882:1995.

[48] Seves A., Beltrame P.L, Selli E., Bergamasco L., *Angew Makromol Chem*, 260; 65–70:1998.

3.15 References

[49] Avella M., Errico, M.E., *J Appl Polym Sci*, 77; 232–236:2000.

[50] Willet J.L., Kotnis M.A., O'Brien G.S., Fanta G. F., Gordon S. H. *J Appl Polym Sci*, 70; 1121–1127:1998.

[51] Shorgen R. L., J Environ *Polym Degrad,* 3(2); 75–80: 1995.

[52] Imam S. H., Chen L., Gordon S. H., Shogren R. L., Weisleader D., Greene R. V., *J Environ Polym Degrad*, 6(2); 91–98:1998.

[53] Imam S. H., Gordon S.H., Shogren R. L., Greene R.V., *J Environ Polym Degrad*, 3; 205–213:1995.

[54] Park H.M., Lee, S. R., Chowdhury S. R., Kong T. K., Park S. H., Ha C. S. *J Appl Polym Sci*, 86; 2907–2915:2002.

[55] Jo Ann Ratto, Peter J. Stenhouse, Margaret Auerbach, John Mitchell and Richard Farrell, *Polymer*, 40(24); 6777–6788:1999

[56] S. Simmons and E L Thomas , *J Appl Polym Sci*, 58; 2259:1995.

[57] S .Simmons and E.L. Thomas, *Polymer*, 39; 5587:1998.

[58] P.J. Stenhouse, J.A. Ratto & N. S Schneider, *J Appl Polym Sci*, 64; 2613:1997.

[59] G.Scott and D. Gilead, In Developments in Polymer Stabilization, Vol. *5,* (Ed. Scott . G) *Appied Science*, London 182;71:1995

[60] G.J.L. Griffin, US Pat. 40, 16, 117 (1997).

[61] K.Seethamraju, M.Bhattacharya, U.R Vaidya and R. G Fulcher, *Rheol Acta*, 33; 553:1994.

[62] U. R. Vaidya and M. Bhattacharya, *J Appl Polym Sci*, 52; 617:1994.

[63] M. Bhattacharya, U. R. Vaidya, D Zhang and R Narayan, *J Appl Polym Sci*, 57; 539:1995.

[64] U. R. Vaidya, M. Bhattacharya and D. Zhang, *Polymer*, 36;1179:1995.

[65] Z. H. Yang and M.Bhattacharya and U. R. Vaidya, *Polymer*, 37; 2137:1996.

[66] G. Scott, In Degradable Polymers Principles and Applications, (Ed. G. Scott and Dan Gilead) *Chapman and Hall*, London, Ch. 9:1995.

[67] X. L. Wang, K. K. Yang and Y. Z. Wang, J. Macromolecular Science, C43;3; 385-409: 2003.

❑❑❑

4

Innovation Till Date

INSIDE THIS CHAPTER

4.0 Research in the field of Biodegradable Polymers as on Date

4.1 References

4.0 RESEARCH IN THE FIELD OF BIODEGRADABLE POLYMERS AS ON DATE

This chapter provides a glimpse of research work carried out by the researchers till date on the development of Biodegradable polymers and their degradation in natural environment like soil burial and compost. Starch is a natural polymer and widely used for the development of Biodegradable polymers. Since it is having properties similar to polymer and is completely biodegradable in its present form, it has been used rigorously by several researchers and has been analyzed through each angle of its reactivity with other degradable or non degradable polymers. Brief of these studies conducted on starch modification, mixing etc. have also been mentioned here.

Aht-Ong et al. [76, 77] have prepared composite films containing various percentages of banana starch, PE-g-MA and benzophenone with LDPE. They have studied the effect of compatibilizer and photosensitizer on composite film through DSC and have analysed the effect on mechanical properties of composite.

Aggarwal et al. [78] has devised a method with a target to measure the removal of starch from the polymer matrix. She has given amylase treatment and followed utilizing thermal analysis.

Albertsson et al. [79] have studied effects of abiotic factors such as water and air on the degradation of polymers poly (3-hydroxybutyrate-co-3-hydroxyvalerate) in simulated and natural composting environments. Evaluated biodegradation rate through Gravimetric analysis, SEM, Size Exclusion Chromatography, DSC, FTIR and NMR.

Ali et al. [80] have prepared biodegradable film from LDPE / sago starch via two-roll mill. They have reported glycerol increases the compatibility and reduce the brittleness of the film. Biodegradation study conducted in fungi environment and evaluated through MFI, tensile strength, percentage elongation. They have also prepared Tapioca starch based packaging films and studied its properties.

Arvanitoyannis et al. [81-84] have prepared biodegradable polymer films with LDPE and rice or potato or wheat starch by extrusion in the presence of varying amounts of water. The presence of high starch contents (30% w/w) have decreased mechanical properties, however gas permeability, water vapour transmission rate and biodegradation rates have increased proportionately with starch contents increase. They have prepared gelatinized starch and 1,4-trans-polyisoprene (gutta percha) composition and compared its properties from packaging and medical application point of view. They have also studied chitosan-poly (vinyl alcohol) system plasticized with sorbitol and sucrose.

Averous et al. [85-90] have reviewed starch based biodegradable multiphase systems and have prepared polysaccharides based composites with the interaction of plasticized starch and cellulose. They have studied the effects of glycerol contents on the starch. These polysaccharides based composites have been compared with polyethylene composites. They have also studied multiple biodegradable systems derived from starch/polyesters, starch/polyesteramide, starch/poly caprolactone.

Bae et al. [91] have analysed the effect of reactive blending on material properties and morphology of HDPE/plasticised starch compositions. HDPE was chemically modified with glycidyl methacrylate in melt blending in the presence of dicumyl peroxide (DCP). A finer dispersion was achieved and miscibility enhanced.

Baltä Calleja et al. [92] have studied evolution of the amorphous structure of starch samples after injection moulding. They studied the influence of the network structure and water content. The results revealed that penetration of water, provokes molecular mobility, hence, observed a better packed helical structure which becomes the precursor of a double helix crystalline formation.

Bastioli et al. [93] has studied the properties and applications of Mater-Bi starch-based materials. Novamont's starch-based biodegradable materials have been classified as A, Z, V, and Y, and were studied in different aspects, such as processability, physico-chemical and physico-mechanical properties, composting behaviour and future market perspectives of Mater-Bi products.

Bellare et al. [94] have prepared strips, from LDPE with well dried, modified, granular starch (CATO-32) through single screw extruder as per Griffin technique. They have analysed degradation of strips in thermal and UV environment through starch hydrolysis by α-amylase at 95°C, thermal oxidized in air oven at 80°C and exposed to 254 nm UV radiations. The results of the study revealed that α-amylase acts on the surface starch to cause cracks, holes, pitting, erosion and through which the starch filled LDPE becomes brittle.

Berslin et al. [95] have studied outdoors the Polyethylene and starch-polyethylene composite films in the straw-line of a marsh and in sea water. They reported that degradation rate is high in the strawline of marsh (attributed to mainly photo degradation) and low in sea water.

Bhattacharya et al. [96-101] have prepared biodegradable polymers from corn starch as well as from high-amylose starch by extrusion with the Ethylene-Vinyl Alcohol Co-polymer. Heterogeneous mixtures of EVAL with yellow corn starch displayed a clearly defined boundary between the phases; the sizes of the domains were proportional to the extrusion time. The water resistance and mechanical properties of starch mixtures are determined by the ratio of their constituents. They have also prepared mixtures of corn starch (25% of amylose, 75 % of amylopectin) with co-polymers containing 18% and 28% of CH_2CHOAC units and about 0.8 % of anhydride fragments by extrusion at different temperatures, screw speeds and mixing times.

Bikiaris et al.[[102-106] have prepared biodegradable polymer samples using Low-density polyethylene and starch in varying starch contents through conventional extrusion, injection-molding, and film-blowing techniques using 0.4 and 0.8% polyethylene-g-maleic anhydride (PE-g-MA) as a compatibilizer. The obtained results show as the amount of anhydride group in the copolymer increase a finer dispersion of starch in the LDPE matrix is achieved. They have also reported that starch, except for being a biodegradable material, can also act as a reinforcing agent and have explained processing-structure-property relationships for reinforcing behaviour of starch. They have studied thermo-oxidation resistance of mixtures of LDPE with a Methyl Methacrylate-Butadiene-Styrene Co-polymer (MBS), with EAA and plasticized starch. A ternary LDPE-EAA-starch system appeared to be the most resistant to temperature, apparently due to the stabilizing effect of the co-polymer at the LDPE-starch interface.

Briassoulis et al. [107] have reviewed the definition of degradability of polymers used in agriculture. Emphasis is placed on the controversial issues regarding biodegradability of some of these polymers.

Bello-Pérez et al. [108-109] have prepared biodegradable polymer films by extruding LDPE with acetylated and oxidized banana starches at different concentrations. They have observed acetylated banana starch based film was smooth but irregularities in native and oxidized banana starches based films. They have analyzed better compatibility in acetylated banana starch based films through DSC. The conducted soil burial biodegradation study for the same. Degradation rate was faster in acetylated banana starch based film.

Calmon-Decriaud et al. [110] have overviewed the standardization activities for biodegradability assessment of polymers and a comparison of the methods used for biodegradability tests on solid polymers and packaging materials.

Canche-Escamilla et al. [111] have conducted enzymatic biodegradation on starch-g-PBA and starch-g-PMMA. Results show that grafted samples with PMMA and PBA achieved degradation of their starch moiety. PBA in starch-g-PBA samples hindered the accessibility to enzymes to the degradable material and resulted long degradation time.

Carvalho et al. [112] have prepared composites with regular cornstarch plasticized with glycerin and reinforced with hydrated kaolin. The kaolin has been used as filler in order to improve mechanical properties of the system. The study concluded that samples filled with 50 phr kaolin showed an increase in the tensile strength and modulus of elasticity and decrease in tensile strain at break.

Cheng G et al. [113] have studied miscibility, crystallization behaviour, tensile properties, and environmental biodegradability of poly (β-hydroxybutyrate) (PHB)/cellulose acetate butyrate (CAB) systems with DSC, SEM, WXRD, and polarizing optical microscopy.

Chang PR et al. [114-115] have conducted water sensitivity study on nanocomposite films which have been prepared from starch/α-zirconium phosphate and mixed suspension of hemp cellulose nanocrystals (HCNs) and thermoplastic starch, or plasticized starch (PS), by the casting and evaporating method.

Chang YH et al. [116] have analysed the effect of acid–alcohol treatment on the molecular structure and physicochemical properties of maize and potato starches. The average granule size of modified starches decreased slightly. The solubility of starches increased with the increase of treatment time, and the pasting properties confirmed the high solubility of modified starches. They have reported degradation of starches occurred mostly within the first 5 days of treatment, and degradation rate of potato starch was higher than maize starch both in amylopectin and in amylose. Maize starch was found less susceptible to acid–alcohol degradation than potato starch.

Christie et al. [117] have done measurement of starch thermal transitions using DSC. They have used maize starches with different amylose content: waxy 0%, corn 27%, G50 50% and G80 80% in the experimental work and have prepared a practical guide to study the thermal behaviour of starch.

4.0 Research in the field of Biodegradable Polymers as on Date

Ciol et al. [118] have studied the effects of extrusion temperature, screw speed and moisture contents on the processing of corn starch using single screw extruder. They reported that starch processing can be effectively controlled by controlling the moisture content of the raw material and the extrusion temperature. They have observed that at high moisture contents of starch, manipulation of screw speed is required while at low moisture content, the processing is independent of screw speed. The extruded starch with low moisture content exhibits a very low retrogradation capacity. It has been observed that variation of operating conditions permit the production of an extrudate with various technological characteristics to meet different industrial applications.

Córdoba et al. [119] have studied the plasticizing effect of alginate on the thermoplastic starch/glycerin systems, Corn starch and corn starch–alginate (5–15%) systems plasticized with 35% glycerin, water was intentionally excluded from the formulations. They have observed when alginate contents increase, elongation at break and impact resistance increase, elastic properties and granular crystalline structure decrease.

Danjaji et al. [120-121] have prepared composites containing various percentages of sago starch and linear low-density polyethylene (LLDPE). It was observed that; tensile strength and elongation at break have decreased; moisture uptake has increased with the increase in starch contents in the composites and moisture barrier properties have decreased at higher relative humidity. In addition they have identified poor wetting between the starch granules and LLDPE matrix. The degradation behaviour has also been studied.

Devi et al. [122] have prepared biodegradable polymers by the modified starch [starch phthalate (upto 30%)] and native starch (upto 30%) in LDPE. They have reported that modified starch gives better mechanical properties and greater degradation rate in soil as compared to LDPE starch system.

Doane et al. [123] have published the effect of moisture contents on tensile properties of starch/poly(hydroxyester ether) (PHEE) composites. They have extruded cornstarch with either 1 or 10% moisture contents with PHEE on total solids basis. They observed that the starch granule structure was progressively disrupted but not completely destroyed as Total Moisture Contents (TMC) increased. Tensile strength and modulus values were not significantly impacted when the TMC was up to 6% and above it decrease rapidly.

Elvira et al. [124] have modified starch chemically to prepare biodegradable polymeric systems and studied the effects on water uptake, degradation behaviour and mechanical properties. Three starch based systems with: (*i*) a copolymer of ethylene and vinyl alcohol (SEVA-C), (*ii*) cellulose acetate (SCA), and (*iii*) poly-e-caprolactone (SPCL); were chemically modified by chain cross-linking. The results show that water uptake of this system could be reduced up to 15% and stiffer material could be obtained.

Evans et al. [125] have examined nanocomposites of glycerol-plasticized starch, with untreated montmorillonite and hectorite and Treated hectorite and kaolinite to produce conventional composites within the same clay volume fraction range. They have reported new class of environmentally accepted biodegradable material which show greater modulus.

4.0 Research in the field of Biodegradable Polymers as on Date

Fabunmi et al. [126] have used starch as fillers in starch-filled polymer systems for the production of foamed starch and biodegradable synthetic polymer.

Favis et al. [127-131] have used a single step method to prepare biodegradable polymer with the help of LDPE/Starch/Glycerol (upto 36%) compositions. Reported excellent tensile properties at 29% and lowest moisture sensitivity at 36% Glycerol. They optimized the composition to (Glycerol/Starch) 30:70 showing excellent results while on further increase in %age dramatic reduction in mechanical properties of the systems. They have analysed particle size of starch and quantitatively compared with average dispersed phase particle size. They proved that one-step processing can be used to generate highly elongated morphological structure. A two-step approach, analogous to typical compounding and shaping operations and involving controlled glycerol removal in the second step can be used to prepare a wide range of highly stable, more isotropic, dispersed particle morphologies. They have also studied binary and ternary systems of (PLA), (PCL) and thermoplastic starch. 50/50 PLA/TPS. They analyzed critical aspects of starch plasticization in glycerol/excess water mixtures in order to determine the time/temperature boundaries ultimately required for the successful plasticization of starch in a polymer melt-processing environment. The efficacy of plasticization is demonstrated by a dramatic six-fold reduction in the dispersed starch phase size.

Fishman et al. [132] have studied micro-structural and thermal dynamic mechanical properties of extruded pectin/starch/glycerol edible and biodegradable films through SEM and DMA.

Study revealed that the temperature profile in the extruder and the amount of water present during extrusion could be used to control the degree gelatinization.

Forssella et al. [133] have studied the effects of glycerol and water content on the thermal transitions of plasticized barley starch using DSC and DMTA. They concluded that low glycerol contents system show single phase while at high contents of glycerol phase separation in the system occurs.

Fowler et al. [134] have reported starch and its derivatives as sources of packaging materials. The result of the study indicated that either starch should be chemically modified or should be combined with polymers that are truly biodegradable into a laminate film that has properties as good as those found in synthetic films. They have reported that these films have better performance than current petrochemical derived materials in terms of gas barrier properties and water resistance.

Ghaemy et al. [135] have defined the grafting procedure of maleic anhydride on polyethylene in a homogeneous medium in the presence of radical initiators (Benzoyl peroxide, azobisisobutyronitrile, and dicumylperoxide).

Gandini et al. [136] have conducted degradation studies on system of conventional cornstarch and glycerol reinforced with cellulosic fibres. The results showed that an increase in glycerol content reduced starch degradation whereas an increase in fibre contents have lead to its increase.

Gao et al. [137] have worked on graft copolymerization of starch-acrylonitrile (AN) initiated by potassium permanganate. They have reported that grafting parameters have significant effect

4.0 Research in the field of Biodegradable Polymers as on Date

on the graft copolymerization and on its components. However, functional group influences the grafting ability of the starch.

Gáspár et al. [138] have prepared biodegradable corn starch based system and PCL based system, containing cellulose, hemi cellulose and zein. They have conducted mechanical properties analysis, water absorption and enzymatic degradation studies. They have reported tensile strength of these samples was significantly higher as compared to PCL filled system. The system thus prepared is biodegradable by enzymes.

Gillmore et al. [139] prepared systems of starch with PP, starch with LDPE, PCL with LDPE, and a copolymer of (PHB/V). They conducted biodegradation studies on these systems in compost. These composts were prepared by municipal leaf composting waste.

Gontard et al. [140] have studied gas transport properties of starch based films. They studied oxygen permeability of plasticised starch, octanoated starch, starch–EVOH system, and PE–starch multilayer. They have reported that the best gas barrier properties of starch based films are generally obtained for low water and plasticiser contents. It was then shown that the interesting low permeability of these systems is due to the low solubility of gases.

Hebeish et al. [141] have demonstrated the basis of novel method for synthesis of poly (methacrylic acid) starch graft copolymers. They have investigated graft copolymerization of methacrylic acid (MAA) on to starch using a potassium persulfate/ sodium thiosulfate redox initiation system.

Ha C-S et al. [142] prepared samples of starch with different thermoplastics. They have reported tensile strength and Modulus increases in starch/ionomer system and decrease in starch/APES, starch/LDPE system with increase in starch contents. Better homogeneity is observed in starch/ionomer systems in comparison to starch/APES and starch/LDPE systems. Up to 50% starch content, the starch/ionomer system appear as a single phase, the extent of phase interactions of starch/APES system lies in between starch/LDPE and starch/ionomer systems.

Hanna et al. [143] have prepared 0.4–0.6 mm thick films of starch with plasticizer (water, glycerol, and stearic acid). They have observed that if glycerol contents increase, increases interactions in between starch and plasticizer. During extrusion in the presence of glycerol, the A-type crystalline structure of starch was transformed to B-type. Crystallinity increased with increase in glycerol content due to tight packing of starch chains.

Herald et al. [144] have investigated the strength of films extruded from corn zein or corn gluten meal (CGM) with LDPE. They have reported that higher the level of biological material (CGM or zein) the lower the tensile properties but CGM containing films exhibited lower tensile properties.

Hirotsu et al. [145] have studied the effect of maleic anhydride grafted polypropylene (MAPP) on the system of cellulose and PP. They have reported MAPP generate strong interaction between PP matrix and cellulose which improved the tensile properties of the system. MAPP containing cellulose composites have high potential for application as environmentally friendly materials.

Ishak et al. [146-149] have prepared sago starch-LLDPE composites and studied its mechanical properties. They have studied the effects of the starch volume fraction, on the tensile modulus and tensile strength through Nicolais and Narkis model. They have determined effects of pro-oxidant (containing transition metals) and unsaturated elastomers (containing sago starch) on the mechanical properties of the film. They have monitored natural weathering effects on SS/LDPE films. They have also studied dry starch filled polycaprolactone and rice starch filled poly lactic acid composite systems.

Jana J et al. [150] has studied the amylose and amylopectin structures of starch and their arrangement. They have also studied various modification methods of starch.

Jagannath et al. [151] have quantified various properties of composition of LDPE with starch 1% and 5% mixing. It is reported that mechanical properties such as %elongation, tensile, tear, bursting and seal strength have decreased and water vapour, oxygen transmission rate increased. However, material still can be used as a packaging material.

Janssen et al. [152] have mixed starch and polyol, as a first step, in a cooking-extrusion process and then as a second step polymer processed through extrusion or film blowing or injection moulding. They prepared films with thicknesses of 200 μm. They demonstrated that the final objects can reach a strength, comparable to commercial packaging plastics, like polystyrene but the problem of moisture sensitivity persists. This problem increase with increasing surface to volume ratio of the final product.

Jinglin et al. [153] have prepared and characterized compatible starch/polyethylene system using maleic anhydride (MAH) through single screw extruder. The thermal plasticization of starch and its compatibilising modification with polyethylene were accomplished. The results showed that maleic anhydride (MAH) could graft onto the polyethylene chain in single step and the systems with MAH have better mechanical properties, thermal stability and compatibility.

Jin et al. [154] have prepared nanoplastics using formamide/ ethanolamine-plasticized thermoplastic starch (FETPS) as the matrix and ethanolamine-activated montmorillonite (EMMT) as the reinforcing phase. They have tested through various testing techniques and observed that FETPS have successfully been intercalated into the layers of EMMT and formed the intercalated nanoplastics with EMMT. On comparison they observed that nanoplastics have effectively restrained the recrystallization of starch and their strength, water resistance and thermal stability has obviously been improved in contrast to pure FETPS.

Jiugao et al. [155-158] have studied hydrogen bond-forming abilities of urea, formamide, acetamide and polyols (glycerol) with starch and the order of bond forming abilities have been defined and confirmed through B3LYP chemical computation. They prepared green composites from corn starch and activated montmorillonite by extrusion. Starch was plasticized with novel plasticizers urea and ethanolamine, and the activated MMT were obtained using ethanolamine as the activated solvent. Results show that mechanical properties were significantly improved. They studied the influence of citric acid on the properties of thermoplastic starch/linear low-density polyethylene blends also.

Jolly et al. [159-160] reviewed the mechanical properties of plasticized starch-based materials from the literature. The methodology relied on the use of a graphic tool allowing a direct comparison of the strength and strain at break. The mechanical properties of the materials were systematically compared with those of glycerol-plasticized starches at 57% relative humidity.

Karlsson et al. [161] prepared a biodegradable polymer film using LDPE and Starch, LDPE and Starch (7.7% with pro-oxident), and a combination of 70% starch and 30% ethylene maleic anhydride. They studied these films in air and water and observed that LDPE with 7.7% starch show the maximum biodegradability.

Kirwan et al. [162] have reported improvement of the impact performance of a starch based biopolymer via the incorporation of Miscanthus giganteus fibres. They examined the possibilities of using an established energy crop; Miscanthus giganteus as a filler/ reinforcement in injection mouldable thermoplastic composites utilising a starch-based biopolymer; Novamont Mater-Bi as the matrix. The addition of Miscanthus fibres to the polymer resulted in composites with higher impact absorbance and loading than those of standard Mater-Bi.

Khonakdar et al. [163-164] have investigated the HDPE and LDPE cross-linking by two methods namely by using different amounts of tert-butyl cumyl peroxide and by irradiation through electron beam beside it they have studied Thermal and mechanical properties of uncross-linked and chemically cross-linked polyethylene/ethylene vinyl acetate copolymer blends. The results obtained suggested that chemically induced cross-linking which took place at melt state, hindered the crystallization process and decreased the degree of crystallinity as well as the size of the crystals which varied proportionally to peroxide contents. While irradiation induced cross linking could not influence the crystalline region. In comparison to HDPE and LDPE, LDPE was more prone to cross linking owing to the presence of tertiary carbon atom and branching as well as owing to its being more amorphous in nature.

Kim H.-J. et al [165] have attempted to prepare Rice Husk Flour (RHF) and Wood Flour (WF) filled polybutylene succinate (PBS) biocomposites as an alternative cellulosic material filled conventional composite and characterised it. They found that as the agro flour loading was increased, the tensile strength and impact strength of the biocomposites decreased. As the filler particle size decreased but the impact strength decreased. The addition of agro flour to PBS produced a more rapid decrease in tensile strength., impact strength, and percentage weight loss of the biocomposite during the natural soil burial test.

Kim M-N et al. [166] have copolymerised ethylene glycol/ adipic acid and 1,4-butanediol/succinic acid in the presence of 1,2-butanediol and 1,2-decanediol to produce ethyl and *n*-octyl branched poly (ethylene adipate) (PEA) and poly (butylene succinate) (PBS), respectively. The chain branching reduced the crystallinity of PEA more significantly than the crystallinity of PBS. Surface tension of PEA was higher than that of PBS, though the two polyesters have identical number of methylene groups and ester groups in the repeating unit. As the degree of chain branching increased, the biodegradation rate of PEA increased to a greater extent than that of PBS due to the faster reduction in the crystallinity of PEA compared to the crystallinity of PBS.

Kiatkamjornwong et al. [167] have chemically modified cassava starch by radiation grafting with acrylic acid to obtain cassava starch graft poly (acrylic acid), which was further modified by esterification and etherification with poly (ethylene glycol) 4000 and propylene oxide, respectively. The modified product was characterized. The composition had much better degradation in soil due to the much higher water absorption.

Kolybaba et al. [168] have reviewed how the modification of traditional material is underway to make them more user-friendly. They have intended to provide a brief outline of work that is under way in the area of biodegradable polymer research and development to design novel polymer composites out of naturally occurring materials.

Kumar et al. [169] have done grafting of maleic acid on LDPE through γ irradiation technique. They used pre-activation method and achieved 2.4% grafting. The grafted polymer (LDPE-g-MAc) was then characterised where presence of high intensity carbonyl peak, high acid value and low melt flow index value confirmed the grafting.

Lee et al. [170] have studied the effects of corn starch (CS) filler and lysine diisocynate (LDI) as a coupling agent on the crystallization behaviour of poly (butylene succinate). Results showed that addition affected the morphological structure of PBS spherulites.

Lescher et al. [171] have done Water-Free mixing of thermoplastic starch and polyethylene for rotomoulding technique.

Liao et al [172] have studied the effects of a compatibilizer on the properties of corn starch reinforced metallocene polyethylene-octene-elastomer (POE) (POE-g-AA) copolymer as a compatibilizer. The result show that POE/Starch system compatibilzed with POE-g-AA copolymer lowered the size of starch phase and had a fine dispersion and homogeneity of starch in the POE matrix. This improved dispersion was done due to formation of branched and cross linked macro molecules because of POE-g-AA copolymer had anhydride groups to react with hydroxyls. This has resulted in improved mechanical properties.

Lindhauer et al. [173] have processed different types of starch with natural plasticizers and commercial fibers. Results show that thermo mechanical treatments of starch caused significant depolymerization, especially of the amylopectin fraction. Tensile strength and water resistance properties were enhanced significantly by the addition of 2–7% fibers.

Liu et al. [174] have studied the effects of glycerin and glycerol monostearate (GMS) on a variety of properties of anhydrous wheat starch. It is found that, as glycerin content (GC) increases, both melting point and degradation temperature decrease, and the range of the processing window extends.

Liircks et al. [175] has studied compostable starch-based plastic material. Starch derivatives and starch esters have been prepared through continuous extrusion process. He has analysed various properties of brands of biodegradable polymer of BIOTEC *i.e.* BIOFLEX, BIOPLAST etc. used for film processing and having properties similar to that of PE.

Lucia et al. [176] have investigated the physical and chemical properties of composite starch-based films containing cellulosic fiber, chitosan, and gelatin. The study revealed that films containing both cellulosic fibers and chitosan have tremendous enhancement in their film strength and gas permeation. Water absorbency in film containing cellulosic fibers and gelatins has greatly reduced, however, by the inclusion of chitosan hydrophilicity and water absorbency increased.

4.0 Research in the field of Biodegradable Polymers as on Date

Maiti et al. [177-178] have prepared starch-based biodegradable (BD) low density polyethylene (LDPE) film and tested the film. This film can be directly printable without any corona treatment, unlike virgin LDPE film. Such a film shows poor adhesion and nail scratch resistance of the ink on the printed area of the film. In order to increase adhesion grafting of acrylonitrile onto the BD film is carried out.

Manzur et al. [179] have determined morphological, structural and surface changes on thermo – oxidized (TO-LDPE) incubated with aspergillus niger and penicillium pinophilum fungi with or without ethanol as substrate. The study revealed that crystallinity and crystalline lamellar thickness decreased upto three units. The higher TO-LDPE changes and fungi-LDPE interaction was observed in the samples with ethanol, suggesting that ethanol favours TO-LDPE degradation.

Ma X.-F. et al. [180-181] have studied properties of Montmorillonite-reinforced thermoplastic starch composites (MTPSC) prepared by melt extrusion with MMT and glycerol-plasticized thermoplastic starch (GTPS). Results confirmed that MMT were uniformly dispersed. They have also worked on wheat flour with the help of urea and formamide and converted into thermoplastic wheat flour (TPF). The study revealed that wheat flour granules were proved to transfer to a continuous phase, lower weight contents of urea and formamide in TPF had better mechanical properties and water resistance.

Mani et al. [182-184] have prepared starch and synthetic polymers based biodegradable polymers. They determined the effect of amylopectin to amylose ratio in the starch on the product.

Mei et al. [185] have prepared LDPE/starch partially biodegradable compounds. They investigated the incorporation of different starches, such as native, adipate, acetylated and cassava starch, in low-density polyethylene matrix. They studied degradation in sludge.

Meng et al. [186] have prepared and studied the properties of poly (propylene carbonate) PPC/Starch composites. Unmodified starch can be simply melt mixed with the PPC to produce biodegradable polymer. Composite properties up to 60% increase in starch contents. The study revealed the interaction between the carboxyl groups of PPC and the hydroxyl group of starch via hydrogen bonding.

Mohammadi et al. [187] have done resistance-to-flow analysis in LDPE/plasticized starch systems containing compatibilizers poly (ethylene-r-vinyl acetate), maleated poly(styrene-ethylene-butadiene-styrene) and maleated polyethylene with Attractive and repulsive interactions. A quasi core/shell structure formation at the dispersed phase/matrix interface was proposed as the origin of resistive capillary flow, leading to a viscoelastic loss enhancement.

Mohanty et al. [188-190] have reviewed the scenario of biodegradable plastics and bio based polymer products. They have suggested that annually renewable agricultural and biomass feedstocks can be the basis of sustainable, eco-efficient products which may takeover the market from petroleum based feedstocks.

4.0 Research in the field of Biodegradable Polymers as on Date

According to them natural bio fiber can effectively replace glass fiber. The combination of bio fibers with renewable and non renewable resources can produce a competitive composite. They suggest that natural fibers with renewable resource based polymers like starch and cellulose can prove to be an efficient alternative to petroleum based composites. Efforts are continuously been made to embed these bio fibers on to renewable materials to achieve the task of energy efficient composites.

Moro et al. [191] have focused on the interactions between plasticized wheat starch matrix and leafwood. They have elaborated different plasticized starch based composites.

Morshedian et al. [192] have studied system of poly (dimethylsiloxane)/low density polyethylene neat and immiscible systems and analysed their interfacial modification.

Müller et al. [193] have presented behaviour of biodegradable polymers prepared with cellulose and thermoplastic and defined its synthesization strategies. The author has distinguished two main groups of cellulose-materials. First one, regenerated cellulose, suitable only for fiber and film production from conventional and new processes point of view and second one, thermo-plastically processable cellulose derivatives such as esters used for extrusion and moulding.

Narayan, et al. [194] have reported drivers for biodegradble/ compostable plastics and role of composting in waste management and sustainable agriculture.

Nayak [195] has reviewed starch based technology and its potential application in biodegradable packaging material.

She has studied tremendous market opportunity from scale business to design new biodegradable material for specific market.

Nho et al. [196] have prepared system of ethylene-1-butene copolymer (EBP) with LDPE to improve the mechanical properties of packaging materials. They have irradiated prepared samples by an electron beam and have examined their thermal and mechanical properties. These properties have significantly been improved.

Nielsen L.E. [197] have described the simple theory of stress-strain properties of filled polymers. By the use of simple models of filled plastics, approximate equations are derived for the elongation to break in the case of perfect adhesion between the phases and for the tensile strength in the case of no adhesion between the polymer and filler phases.

Nikazar et al. [198] have studied corn starch-LDPE films containing starch in different ranges with oleic acid, maleic anhydride and with benzoyl peroxide. Have studied these for fungal growth simulation. Results are dependent on the starch ratio.

Ohtake et al. [199] have examined the molecular weight reduction of low-density polyethylene (LDPE) buried under bioactive soil for 32-37 years. Results show that the formation of the low-molecular-weight components is the result of erosion due to enzymatic reaction. The results give evidence for the biodegradation of high-molecular-weight LDPE.

Otey et al. [200] have started this revolution way back and have prepared foam from starch derived polyethers, starch filled PVC Plastics and studied starch reactivity with polyols.

4.0 Research in the field of Biodegradable Polymers as on Date

Panayiotou et al. [201-204] have prepared biodegradable polymers using modified starch (octanoated starch OCST by esterification of native starch with octanoyl chloride) and plasticised starch, as different samples with low density polyethylene in various proportions. Its biodegradation was followed in activated sludge from a waste water treatment plant. From weight loss during the biodegradation period, it was found that OCST, even with such a high degree of substitution (2.7), is biodegradable. Soil burial degradation study conducted on OCST/LDPE biodegradable polymer for 6 months. High starch composition system showed high reduction in mechanical properties during the biodegradation period.

Park H.-R. et al. [205] have examined the effects of additives with different functional groups *i.e.* hydroxyl and carboxyl groups, on the physical properties of starch/PVA composition. Starch/PVA films have been prepared. Glycerol (GL) with 3 hydroxyl groups, succinic acid (SA) with 2 carboxyl groups, malic acid (MA) with 1 hydroxyl and 2 carboxyl groups, and tartaric acid (TA) with 2 hydroxyl and 2 carboxyl groups were used as additives. The results of measured tensile strength and elongation verified that hydroxyl and carboxyl groups as functional groups increased the flexibility and strength of the film.

Park H.-R. et al. [206] have prepared Starch/polyvinyl alcohol (PVA) films by using corn starch, polyvinyl alcohol (PVA), glycerol (GL), and citric acid (CA) as additives and glutaraldehyde (GLU) as cross-linking agent for the mixing process. The additives, drying temperature, and the influence of cross-linker of films on the properties of the films were investigated. At all measurement results, except for degree of swelling, the film adding CA was better than GL because hydrogen bonding at the presence of CA with hydroxyl group and carboxyl group increased the inter/intramolecular interaction between starch, PVA, and additives.

Park H.-J et al. [207] have produced Biodegradable plastics from sweet potato pulp (SPP) and cationic starch (CS) or chitosan composite (CC) by compression molding and tested their mechanical properties.

Pandey et al. [208-209] have demonstrated that the material properties of polymers can be enhanced dramatically by incorporating layered silicates at fairly low concentrations. They studied biodegradability of UV-irradiated films of ethylene-propylene copolymers (E-P copolymer), isotactic polypropylene (i-PP), and LDPE in composting and culture environments.

Plackett, D. et al. [210] have investigated the inclusion of biodegradable polymers in composites. In their research they have incorporated lignocellulosic plant fibers in biodegradable polyester matrixes and found positive results in terms of composite mechanical properties.

Poovarodom1 et al. [211] have done modification of Casava Starch by esterification and studied Cassava Starch Ester Films Properties. The effects of the ester group chain-length and the degree of substitution (DS) on the properties of starch ester films were studied. The results indicate that starch esterification can improve the hydrophobicity of starch ester films.

Ribes-Greus et al. [212] have mixed three commercial biodegradable additives (Mater-Bi, Cornplast, and Bioefect) into LDPE and prepared LDPE samples filled with these additives. These samples were degraded by soil burial testing. The study revealed that LDPE-Mater-Bi exhibited faster changes in their crystalline content. However, the LDPE samples with cornplast and Bioefect displays more significant changes in their lamellar thickness distribution.

Raghvan et al. [213] have examined the microstructural aspects of vernonia oil-added starch–polyethylene and vernonia oil added starch–polyethylene–polylactic acid composite. Polymer composite films containing different percentage of additives were melt processed and acid hydrolyzed. Results showed that the vernonia oil is present at the interface of the starch–polyethylene.

Rajulu et al. [214] have studied the mechanical, optical and electrical properties of low density polyethylene (LDPE) and linear low density polyethylene (LLDPE) films with and without fillers. The variation of properties resulting from the addition of inorganic fillers included an increase in haze, diffusion of light, and tear strength for films made with both LDPE and LLDPE.

Ratto et al. [215] have investigated composites of a biodegradable thermoplastic aliphatic polyester, polybutylene succinate adipate (PBSA), with granular corn starch. The PBSA/starch films were prepared with various starch contents (5–30%) and processed by blown film extrusion. Increasing the starch content, led to an increase in modulus and decreases in tensile strength, elongation to break and toughness. The rate of biodegradation in soil increased significantly as the starch content was increased to 20%.

Ren, et al. [216] has reviewed the biodegradable plastics. He suggested that though this is seen as a solution for the waste problem, biodegradable plastics create new challenges on waste management with respect to policies and laws, waste management technologies and application of market-based instruments.

Rivard et al. [217] prepared novel series of starch esters and varied both in degree of substitution and ester group chain length (C-2 to C-6). Then measured the effects of starch modification on anaerobic biodegradation potential for the polymer using biochemical methane (BMP) potential protocol.

Rosa et al. [218] have prepared recycled LDPE/corn starch compositions for comparison along-with virgin LDPE/corn starch. Study revealed that the interfacial interaction was weak for compositions containing virgin and recycled LDPE.

Sales et al. [219] have studied Degradation of different polystyrene/thermoplastic starch blends buried in soil. Starch and polystyrene systems prepared using two different plasticizers: glycerol or buriti oil by solvent casting technique. Under soil burial presence of starch at contents of 50% or greater improved the degradation of the systems. Results revealed that Polystyrene's degradability can be improved when starch plasticized with buriti oil is added to it.

Shah PB et al. [220] have characterized the initial degradation mechanism of starch filled LDPE. LDPE was compounded with well dried, modified, granular starch (CATO-32) according to the Griffin technique. The results showed that due to starch hydrolysis α-amylase acts on the surface starch to cause cracks, holes, pitting and erosion. The starch filled LDPE becomes brittle when it undergoes thermal oxidation.

Rao et al. [221] investigated PE-Starch system with or without vegetable oil as compatibilizer. They observed that vegetable oil as an addition has a dual role: as plasticizer it improves the film quality; as a pro-oxidant, it accelerates degradation of film.

Sailaja et al. [222-224] have studied LDPE and plasticized starch using poly (ethylene-co-vinyl alcohol) (EVOH) as a compatibilizer and Itaconic acid-grafted as compatibilizer for LDPE-plasticized Tapioca Starch. They studied the effect of maleate ester mixing also.

Siddaramaiah et al. [225-226] have prepared a series of LDPE films filled with different fillers such as silica, mica, soya, protein, isolate potassium permanganate, and alumina using single screw extruder. Results show that due to reinforcing effect induced by the fillers like silica, mica, the tensile strength has improved. Potassium permanganate filled LDPE samples showed significant increase in percentage elongation.

Somwangthanaroj et al. [227] have prepared blown film by LDPE clay from packaging application point of view. The Mechanical and Gas Barrier Properties of Blown Film have been studied by them.

Suh et al. [228] have modified starch by the introduction of cholesterol unit and then prepared different starch-compositioned high-density polyethlene (HDPE) films with native starch/modified starch. They observed that HDPE-blown films containing more than 10% native starch (HDPE/ST) showed a steeper decrease in crystallinity and results indicate that the modified starch has improved dispersion and adhesion with HDPE.

Swift et al. [229] have reported that film samples of LDPE-TDPA™ (pro-oxidant additives) were biodegradable by soil microorganisms with long incubation time. A positive biodegradation profile shows that the bio-degradation continues and changes the structural characteristics of samples.

Tharanathan et al. [230] have reviewed the biodegradable films and composite coatings: past, present and future. He has analysed the naturally occurring polymers and polymeric material used in the film production and its coating possibilities with natural polymers cellulose etc.

Truter et al. [231] have developed biodegradable polymers films by trans-esterification of thermoplastic starch (prepared using wheat starch) with poly (vinyl acetate) and poly (vinyl acetate-co-butyl acrylate) in the presence of sodium carbonate, zinc acetate and titanium (IV) butoxide. The prepared polymer appeared as homogeneous, translucent films with one glass transition temperature range, between that of starch and of the polymer. The presence of wheat starch in the samples improved the mechanical strength.

US Patent [232] Disclosed herein is a composition of matter comprising a blend of discontinuous thermoplastic starch domains in a synthetic polymer, said composition of matter being characterized by an average diameter of thermoplastic starch domains of about 0.2 to about 1.5 microns. Also disclosed are compositions of matter comprising a blend of discontinuous thermoplastic starch domains in a synthetic polymer and being characterized by finished articles having key mechanical properties which are essentially maintained or in some cases improved over pure synthetic polymers. In yet another aspect, the present invention provides a method for making the material of the present invention. In a related aspect, the present invention provides the novel materials issued from the method of making the material. In

4.0 Research in the field of Biodegradable Polymers as on Date

other aspects, the present invention provides novel finished article compositions in the form of films or molded articles.

Vaidya et al. [233-237] have prepaerd the mixtures of starch with an Ethylene-Propylene Co-polymer (EP), polystyrene (PS), an Ethylene-Propylene-Maleic Anhydride Co-polymer (0.8%, EPMA) as well as with a Styrene-Maleic Anhydride Co-polymer (8%, SMA) and studied at different effective mechanical moulding energies. They tested biodegradability of the system also.

Vašková et al. [238] prepared multicomponent systems based on three main components: polycaprolactone, native corn starch and polyhydroxybutyrate. Two-components system composed of polycaprolactone (PCL) and native corn starch (as master system) have been modified by the addition of polyhydroxybutyrate (PHB). These mixtures were prepared using twin screw extruder by two different techniques one step and two step. Two-step process provides better incorporation of individual plasticizers into reasonable part of polymer systems. Moreover, a better dispersion can be achieved in comparison with one step process.

Verhoogt et al. [239] have studied starch as the dispersed component in a polyethylene (LDPE or LLDPE) matrix in the form of dry granular starch and gelatinized **starch** plasticized with glycerol. Study indicates the possibility of achieving a level of morphological control with respect to the size and shape of the dispersed phase in case of gelatinized starch. Results show that at high loading of gelatinized starch has improved the mechanical properties of the system.

Vinhas et al. [240-241] have presented a study of the effect of the distinct starch types on both the growth and the amylolitic activity of mixed and isolated cultures of the fungi.

Vliegenthart et al. [242-244] have studied low and high molecular mass **thermoplastic starch.** The granular **starches** were plasticized by extrusion processing with glycerol and water. The low molecular mass **starch** was prepared by partial acid hydrolysis of potato **starch.** Results show that the stress-strain properties of the materials were dependent on the water content. They studied influence of crystallization on the stress-strain behaviour of potato **starch.** They have determined through experiments that B type crystallinity can be increased through the processing of starch with Glycerol and it improves the mechanical properties in comparison to native starch.

Walia et al. [245] have studied the properties of thermoplastic starch and poly (hydroxy ester ether) system as a function of the starch concentration. They show that the moisture level present during processing significantly affects the morphology. They conducted morphology study. The system modulus could be effectively represented by the particular morphology present at any given starch concentration range, using a generalized form of Kerner's equation.

Wang Y.Z. et al. [246] have reviewed starch based totally biodegradable polymers which are comparable with non degradable polymers. They have studied mechanical properties of such system which are close to traditional plastic such as polyethylene.

Wang Y. J. et al. [247] have measured the effects of PE-g-MA on the samples of LDPE and corn starch. The observations are that the crystallization temperature of LDPE – corn starch / PE-g -MA samples was similar to LDPE but higher then that of LDPE-corn starch system. The interfacial properties between corn starch and LDPE improved. Tensile strength and elongation at break improves in high amount of starch.

Wilhelm et al. [248] have used mineral clay as filler in order to improve the mechanical properties of glycerol-plasticized Cará starch films. These prepared system has been characterized. They shoed that Clay exfoliation occurred in unplasticized starch/clay mixtures.

Willett et al. [249] have presented tensile properties of granular starch-filled low-density polyethylene (PE) and starch-filled system of PE and poly (hydroxy ester ether) (PHEE). They described properties using simplified form of the Kerner equation.

Williams PA et al. [250, 251] have done Physico-chemical characterization and rheological properties of sago starch.

Wollerdorfer Martina et al. [252] have prepared biodegradable polymers using polyesters, polysaccharides and thermoplastic starch. Then, they have studied the influence of reinforcing of plant fibers on the mechanical properties of biodegradable polymers. They determined that the chemical similarity of polysaccharides and plant fibers has resulted in an increased tensile strength of the reinforced polymers. For reinforced thermoplastic wheat starch, it was four times better than without fibers.

Wool et al. [253] studied degradation of polyethylene-starch system in soil. They exposed binary polymer films containing different percentages of corn starch and low-density polyethylene (LDPE) to soils over a period of 8 months and monitored for starch removal and chemical changes of the matrix. Weight loss data were consistent and suggested that starch removal continues past 240 days.

Wootthikanokkhan et al. [254] have done the modification of cassava starch by using propionic anhydride and studied the properties of the starch-blended polyester polyurethane. The reaction was carried out in the presence of a pyridine catalyst and the effects of reaction variables such as the anhydride content, reaction time and reaction temperature on the degree of acylation were investigated. Obtained results show better interfacial adhesion than the normal starch-blended polyurethane.

Wu C. S. [255] have studied the properties of samples of polycaprolactone and starch (PCL/starch) and maleic anhydride (MAH)-grafted-polycaprolactone and starch (PCL-g-MAH/ starch). The greater compatibility of PCL-g-MAH with starch, owing to the formation of an ester carbonyl group, led to a much better dispersion and homogeneity of starch in the PCL-g-MAH matrix and consequently to noticeably better properties. Furthermore, with a lower melt temperature, the PCL-g-MAH/ starch system is more easily processed than PCL/starch. Both systems were biodegradable.

Xiaodong et al. [256] have synthesized a graft, oxidation starch sizing agent. Analyzed its structure and surface morphology. Result show that synthesized graft oxidation starch has good desizing ability and lower hygroscopic properties. In another study synthesized carboxymethyl starch (CMS) with a high degree of substitution (0.91) and high viscosity by optimized reaction conditions. Studied its structure and surface morphology. Results show that CMS exhibited excellent printing behaviour.

Xiong et al. [257] have prepared starch-based biodegradable film, with a nano silicon dioxide (nano-SiO_2), by the coating method, through which its physical and biodegradable properties were studied. The structure of the film was characterized by Fourier transform infrared spectroscopy (FT-IR), X-ray photoelectron spectroscopy (XPS), X-ray diffraction (XRD), differential scanning calorimeter (DSC), and scanning electron microscopy (SEM). The obtained results are: crystallinity decreased, tensile strength, breaking elongation and transmittance have increased.

Yoon J. S. et al. [258] have prepared polypropylene (PP)/ montmorillonite (MMT) nanocomposites by esterification of propylene-g-maleic anhydride (MAPP) with MMT modified with α,ω-hydroxy amine. Results show that MMT retarded crystallization of MAPP. Compounding PP with MAPP/MMT composites enhanced the tensile modulus and tensile strength of PP. However, the Elongation at break decreased drastically even when the MMT control was very low in content.

Yoshii et al. [259] have studied the influence of radiation, plasticizers, water and poly vinyl alcohol (PVA) on the properties of the sheets. They prepared transparent starch-based plastic sheets by irradiation of compression-molded starch-based mixture in physical gel state with electron beam (EB) at room temperature. Glycerol, ethylene glycol (EG), poly ethylene glycol (PEG, 600, 1000) was selected as plasticizer to add into starch sheets. The results showed that **glycerol was an excellent plasticizer** of starch so that the ductility of starch sheets was improved obviously.

Zainuddin et al. [260] have evaluated the biodegradability of Binolle and compatibilized CPP/Binolle system in two biotic environments, which is soil and lipase enzyme solution. The results show that mechanical properties of samples reduced significantly in the soil burial. During study it has been observed that both amorphous and crystalline parts were randomly attacked.

Zao et al. [261] have mixed starch with glycerol and urea and then extruded in twin screw extruder. The observations of study are, in general tensile strength has reduced and percentage elongation increased as glycerol acts as plasticizer and urea breakdown interaction among starch molecules.

Zobel et al. [262] has reviewed the structure of starch right from molecule to granules.

4.1 REFERENCES

[76] Usarat Ratanakamnuan, Duangdao Aht-Ong, *J Appl Poly Sci*, 100(4); 2717–2724:2006.

[77] Usarat Ratanakamnuan, Duangdao Aht-Ong, *J Appl Polym Sci*, 100(4); 2725–2736:2006.

[78] Poonam Aggarwal, *Thermo Acta* 340–341;195–203:1999.

[79] Carina EldsaÈter, Sigbritt Karlsson, Ann-Christine Albertsson, *Polym Degrad Stab*, 64; 177–183: 1999.

[80] R. Rasit Ali, W. A. Wan Abdul Rahman, N. Zakeria., International Conference on Advancement of Materials and Nanotechnology – The City Bayview Hotel, Langkawi, Kedah, Malaysia. 112:2007.

[81] Ioannis Arvanitoyannis, Costas G. Biliaderis, Hiromasa Ogawa and Norioki Kawasaki, *Carbo Polym*, 36(2-3); 89–104:1998.

[82] Ioannis Arvanitoyannis, Eleni Psomiadou, Costas G. Biliaderis, Hiromasa Ogawa and Norioki Kawasaki, *Carbo Polym*, 33(4); 227–242:1997.

4.1 References

[83] Ioannis Arvanitoyannis, Ioannis Kolokuris, Atsuyoshi Nakayama and Sei-ichi Aiba, *Carbo Polym*, 34(4); 291–307:1997.

[84] Ioannis Arvanitoyannis, Ioannis Kolokuris, Atsuyoshi Nakayama, Noboru Yamamoto and Sei-ichi Aiba, *Carbo Polym*, 34(1-2); 9–19:1997.

[85] Averous L, Boquillon N, *Carbo Polym*, 56; 111–122:2004.

[86] Averous L, *J Macrom Sci Polym R C*, 44; 231–274:2004.

[87] Averous L, Fringant C, Moro L, *Polymer*, 42; 6565–6572:2001.

[88] Averous L, Fringant C, *Polymer Eng Sci*, 41; 727–734:2001.

[89] Averous L, Moro L, Dolo P, Fringant C, *Polymer*, 41; 4157–4167:2000.

[90] Averous L, Fringant C, Fauconnier N, Moro L, *J Appl Polym Sci*, 76; 1117–1128:2000.

[91] Byung Chul Jang, Soo Young Huh, Jeong Gyu Jang, Young Chan Bae, *J Appl Polym Sci*, 82(13); 3313–3320: 2001.

[92] R. K. Bayer, M.E. Cagiao, F. J. Baltä Calleja, *J Appl Polym Sci*, 99; 1880–1886:2006.

[93] Catia Bastioli, *Polym Degrad Stab*, 59; 263–212:1998.

[94] Piyush B. Shah, S. Bandopadhyay and Jayesh R. Bellare, *Polym Degrad Stab*, 47;165–173:1995.

[95] Vincent T. Breslin, Boen Li, *J Appl Polym Sci*, 48(12); 2063–2079:2003.

[96] Mani, R., Bhattacharya, M., *Eur Polym J*, 37; 515–526:2000.

[97] M. Bhattacharya, R.S. Singhal and P.R. Kulkarni, *Carbo Polym*, 31;79:1996.

[98] U.R. Vaidya, M. Bhattacharya, D. H. S. Ramkumar, M. Hakkarainen, A.C. Albertsson, and S. Karlsson, *Eur Polym J*, 32; 999:1996.

[99] D.H.S. Ramkumar, M. Bhattacharya and U.R. Vaidya, *Eur Polym J*, 33;729:1997.

[100] D.H.S. Ramkumar, Z.Yang and M. Bhattacharya, *Polym Networks Blends*, 7; 31:1997.

[101] D.H.S. Ramkumar and M Bhattacharya, *J Mater Sci*, 32 (1997), p. 2565

[102] D. Bikiaris and C. Panayiotou, *J Appl Polym Sci*, 70(8); 1503–1521:1998.

[103] C. Panayiotou, P. Matzinos, D. Bikiaris, S. Kokkou, *J Appl Polym Sci*, 79(14); 2548–2557:2001.

[104] D. Bikiaris, J. Prinos, and C. Panayiotou, *Polym Degrad Stab*, 57;1:1997.

[105] D. Bikiaris, J. Prinos, C. Perrier and C. Panayiotou, *Polym Degrad Stab*, 57; 313:1997.

[106] D. Bikiaris, J. Prinos and C. Panayiotou, *Polym Degrad Stab*, 58; 215:1997.

[107] Ioanna Kyrikou and Demetres Briassoulis, *J Polym Envir on*, 15(2); 125–150:2007.

[108] Arturo Bello-Pérez, *J Appl Polym Sci*, 106(6);3994–3999.

[109] Apolonio Vargas Torres, Paul Baruk Zamudio-Flores, René Salgado-Delgado, Luís Arturo Bello-Pérez, *J Appl Polym Sci*, 110(6);3464–3472:2008.

[110] Anne Calmon-Decriaud, Véronique Bellon-Maurel and Françoise Silvestre, *Advan Polym Sci*, 135; 207–226:1998.

[111] M.V. Moreno-Chulim, F. Barahona-Perez, G. Canche-Escamilla, *J Appl Polym Sci*, 89; 2764–2770:2003.

[112] A. J. F. de Carvalho, A. A. S. Curvelo and J. A. M. Agnelli, *Carbo Polym,* 45(2); 189–194: 2001.

[113] Tiezhu Wang, Guoxiang Cheng, Shihu Ma, Zhijiang Cai, Liguang Zhang, *J Appl Polym Sci,* 89; 2116–2122: 2003.

[114] Xiaodong Cao, Yun Chen, Peter R. Chang, Mark Stumborg, Michel A. Huneault, *J Appl Polym Sci,* 109(6); 3804–3810: 2008.

[115] Haixia Wu, Changhua Liu, Jianguang Chen, Peter R. Chang, Yun Chen and Debbie P. Anderson, Article in Press, Corrected Proof, *Carbo Polym*, accepted 7 January, 2009.

[116] Jheng-Hua Lin, Shyh-Yang Lee, Yung-Ho Chang, *Carbo Polym* 53; 475–482: 2003.

[117] L. Yu and G. Christie, *Carbo Polym,* 46(2); 179–184: 2001.

[118] Ahmed A. El-Dash, Rolando Gonzales and Marcia Ciol, *J Food Eng,* 2(2); 129–152: 1983.

[119] Andrés Córdoba, Nicolás Cuéllar, Mauricio González and Jorge Medina,*Carbo Polym*, 73(3); 409–416:2008.

[120] I. D. Danjaji, R. Nawang, U. S. Ishiaku, H. Ismail, Z. A. Mohd. Ishak, *J. Appl. Polym. Sci.,*79(1); 29–37:2001.

[121] I. D. Danjaji, R. Nawang, U. S. Ishiaku, H. Ismail, Z. A. Mohd. Ishak, *Polym. Test* , 21; 75–81: 2002.

[122] I. M. Thakore, Sonal Desai, B. D. Sarawade and Surekha Devi, *Euro Poly J* , 37(1); 151–160:2001.

[123] J. L. Willett and W. M. Doane, *Polymer,* 43(16); 4413–4420: 2002.

[124] Döne Demirgöz, Carlos Elvira, Joäo F. Mano, Antonio M. Cunha, Erhan Piskin, Rui L. Reis, *Polym Degrad Stab,* 70; 161–170:2000.

[125] Biqiong Chen and Julian R.G. Evans, *Carbo Polym*, 61(4); 455–463:2005.

[126] Olayide Oyeyemi Fabunmi, Lope G. Tabil, Peter R. Chang, Satyanarayan Panigrahi, Paper number RRV07130, ASABE/CSBE North Central Intersectional Meeting, 2007.

[127] F. J. Rodriguez-Gonzalez, B. A. Ramsay and B. D. Favis, *Polymer,* 44(5), 1517–1526: 2003.

[128] F. J. Rodriguez-Gonzalez, B. A. Ramsay and B. D. Favis, *Carbo Polym,* 58(2); 139–147:2004.

[129] F. J. Rodriguez-Gonzalez, N. Virgilio, B. A. Ramsay, B. D. Favis, *Adv Polym Techno*, 22(4); 297–305:2003.

[130] Pierre Sarazin, Gang Li, William J. Orts and Basil D. Favis, *Polymer*,49(2); 599–609:2008.

[131] Gang Li, Pierre Sarazin, Basil D. Favis, *Macromol Chem Phys*, 209(10); 991–1002:2008.

[132] M. L. Fishman, D. R. Coffin, R. P. Konstance and C. I. Onwulata, *Carb Polym*, 41(4); 317–325:2000.

[133] Pirkko M. Forssella, Jaana M. Mikkilti, Graham K. Moates b and Roger Parker b, *Carbo Polym,* 34; 275–282:1997.

[134] Jianmin Fang and Paul Fowler, *Food, Agri and Env*, 1(3&4); 82–84: 2003.

[135] Mousa Ghaemy, Solaiman Roohina, *Iran Polym J*, 12(1); 21–29:2003.

[136] A. J. F. Carvalho, M. D. Zambon, A.A.S. Curvelo and A. Gandini, *Polym Degrad Stab,* 79(1); 133–138:2003.

[137] Jianping Gao, Jiugao Yu, Wei Wang, Liming Chang, Ruchuan Tian, *J Appl Poly Sci*, 68; 1965–1972:1998.

[138] M. Gáspár, Zs. Benkõ, G. Dogossy, K. Réczey, T. Czigány, *Polym Degrad Stab*, 90; 563–569:2005.

[139] David F. Gilmore, S. Antoun, Robert W. Lenz, Steve Goodwin, Richard Austin and R. Clinton Full, *J Indu Micr ob Biotechno*, 10(3-4);199–206:1992.

[140] P. Dole, C. Joly, E. Espuche, I. Alric and N. Gontard, *Carbo Polym,* 58(3); 335–343:2004.

[141] A. Hebeish, M. K. Beliakova, A. Bayazeed, *J Appl Polym Sci*, 68; 1709–1715: 1998.

[142] Hwan-Man Park, Sang-Rock Lee, Subhendu R. Chowdhury, Tae-Kyu Kang, Hak-Kil Kim, Seung-Hoon Park, Chang-Sik Ha, *J Appl Polym Sci*, 86(11); 2907–2915, 2002.

[143] Heartwin A. Pushpadass, David B. Marx, Randy L. Wehling, and Milford A. Hanna, *Cereal Chem*, 86(1); 44–51: 2009.

[144] Thomas J. Herald, Ersel Obuz, Wesley W. Twombly, and Kent D. Rausch, *Cereal Chem*,79(2); 261–264:2002.

[145] Wulin Qiu, Farao Zhang, Takashi Endo, Takahiro Hirotsu, *J Appl Polym Sci*, 87; 337–345:2003.

[146] Yew GH, Mohd Yusor AM, Ishiaku US, Mohd Ishak ZA, *Polym Degr Stab*, 90; 488–500:2005.

[147] Danaji ID, Nawang R, Ishiaku UL, Ismail H, Mohd Ishak ZA, *Polym Test* 21; 75–81:2002.

[148] Ishiaku US, Pank KW, Lee WS, Mohd Ishak ZA, Euro *Polym J*, 38; 393–401:2002.

[149] Sharma N, Chang LP, Chu YL, Ismail H, Ishiaku US, Mohd Ishak ZA,*Polym Degr Stb*, 71; 381–393:2001.

[150] Jana J. *J M S Pure Appl Chem*, A32(4); 751–757:1995.

[151] J. H. Jagannath, S. Nadanasabapathi, A. S. Bawa, *J Appl Polym Sci*, 99, 3355–3364:2006.

[152] Leon P.B.M. Janssen, Leszek Mo_cicki ,*Acta Sci Pol, Technica Agraria,* 5(1); 19–25:2006.

[153] Wang Shujun, Yu Jiugao and Yu Jinglin, *Polym Degrad Stab,* 87(3); 395–401:2005.

[154] Ming-fu Huang, Jiu-gao Yu, Xiao-fei Ma and Peng Jin, *Polymer*, 46(9); 3157–3162: 2005.

[155] Jiugao Yu, Tong Lin, Jianping Gao, *J Appl Polym Sci,* 62(9); 1491–1494: 1998.

[156] Xiaofei Ma and Jiugao Yu, *Carbo Polym,* 57(2); 197–203: 2004.

[157] Mingfu Huang, Jiugao Yu, *J Appl Polym Sci*, 99; 170–176:2006.

[158] Wang Ning, Yu Jiugao, Ma Xiaofei and Wu Ying, *Carbo. Polym*, 67(3); 446–453:2007.

[159] N. Follain, C. Joly , P. Dole , C. Bliard, *J Appl Polym Sci*, 97(5); 1783–1794:2005.

[160] N. Follain, C. Joly, P. Dole, B. Roge, M. Mathlouthi, *Carbo Polym*, 63; 400–407:2006.

[161] Minna Hakkarainen, Ann-Christine Albertsson, Sigbritt Karlsson, *J Appl Polym Sci*, 66; 959–967:1997.

[162] Mark Johnson, Nick Tucker, Stuart Barnes and Kerry Kirwan, *Indus Crops Products, 2004.*

[163] H. A. Khonakdar, S. H. Jafari, M. Taheri, U. Wagenkenecht, D. Jehnichen, L. Häussler, *J Appl Polym Sci*, 100; 3264–3271:2006.

[164] H. A. Khonakdar, S. H. Jafari, A. Haghighi-Asl, U. Wagenknecht, L. Häussler, U. Reuter, *J Appl Polym Sci* 103; 3261–3270: 2007.

[165] Hee-Soo Kim, Han-Seung Yang, Hyun-Joong Kim, *J Appl Polym Sci.* 97; 1513–1521: 2005.

[166] Mal-Nam Kim, Keun-Hwa Kim, Hyoung-Joon Jin, Jong-Kyu Park and Jin-San Yoon, *Euro Polym J*, 37(9); 1843–1847,2001.

[167] Suda Kiatkamjornwong, Prodepan Thakeow and Manit Sonsuk, *Polym Degrad Stab*, 73(2); 363–375:2001.

[168] M. Kolybaba, L.G. Tabil, S. Panigrahi, W.J. Crerar, T. Powell, B. Wang, *Paper : RRV03-0007,* CSAE/ASAE Annual Intersectional Meeting, October 3–4, 2003.

[169] S. K. Singh, S. P. Tambe, S. B. Samui, Dhirendra Kumar, *J Appl Polym Sci.* 93; 2802–2807: 2004.

[170] Tmtomu Okhita, Seung-Hwan Lee, *J Appl Polym Sci* , 97; 1107–1114: 2005.

[171] Peter Lescher, Krishnan Jayaraman, Debes Bhattacharyya, *Starch-Starke*, 61(1); 43–45, 2008.

[172] Chin-San Wu, Hsin-Tzu Liao, *J Appl Polym Sci.* 86; 1792–1798: 2002.

[173] U. Funke, W. Bergthaller, and M.G. Lindhauer, *Polym Degrad Stab*, 59; 293–296:1998.

[174] Z. Q. Liu, X.-S. Yi and Y., Feng, *J Mat Sci* , 36(7); 1809–1815:2001.

[175] Jiirgen Liircks, *Polym Degrad Stab* 59; 245–249:1998.

[176] Weiping Ban, JianguoSong, Dimitris S. Argyropoulos, Lucian A. Lucia, *J Appl Polym Sc*i, 100; 2543–2548: 2006.

[177] T Jana, BC. Roy, R Ghosh, S Maiti, *J Appl Polym Sci*, 79; 1273–1277:2001.

[178] Jana TT, Maiti SK, *Macromol Chem*, 267; 16–19:1999.

[179] T. Volke-Sepulveda, G. Saucedo-Castaneda, M. Gutierrez-Rojas, A. Manzur., E. Favela-Torres, *J Appl Polym Sci*, 83; 305–314: 2002.

[180] M.-F. Huang, Jiu-Gao Yu and Xiao-Fei Ma, *Polymer*, 45(20); 7017–7023: 2004.

[181] XF. Ma, JG. Yu and YB. Ma., *Carbo Polym*, 60(1); 111–116:2005.

[182] Mani R, Bhattacharya M, *Euro Polym J*, 37; 515–526:2000.

[183] Mani R, Bhattacharya M, *Euro Polym J*, 34; 1477–1487:1998.

[184] Mani R, Bhattacharya M, *Euro Polym J*, 34; 1467–1475:1998.

[185] E.M. Nakamura, L. Cordi, G.S.G. Almeida, N. Duran and L.H.I. Mei, *J Mat Proces Tech*, 162–163; 236–241:2005.

[186] XC GE, XH Li, Q ZHU, L Li and Meng YZ, *Polym Eng Sci*, 44(11); 2134:2004.

[187] Morteza Ghafoori, Naser Mohammadi, and Seyyed Reza Ghaffarian, *Iranian Polym J*, 15 (9); 747–755:2006.

[188] A. K. Mohanty, M. Misra, and L. T. Drzal, *J Polym Environ*, 10(1): 19:2002.

[189] A. K. Mohanty, M. Misra, *Polym Plast T echno Eng*, 34; 729–792:1995.

[190] A. K. Mohanty, M. Misra, and L. T. Drzal, *Compos Interf,* 8: 313–343:2001.

[191] L. Avérous, C. Fringant and L. Moro, *Polymer,* 42(15); 6565–6572: 2001.

[192] Farhad Faghihi, Jalil Morshedian, Mohamad Razavi-Nouri and Morteza Ehsani, *Ind Polym J*, 17(10); 755–765:2008.

[193] J. Simon, H. P. Müller, R. Koch and V. Müller, *Polym Degrad Stab.,* 59(1-3); 107–115:1998.

[194] Narayan R., *Orbit J,* 1(1); 1–9: 2001.

[195] Nayak L Padma, *J Macromol Chem Phys,* C39(3), 481–505:1999.

[196] Young Chang Nho†, Jeong Il Kim, and Phil Hyun Kang, *J Ind Eng Chem* 12(6), 888-892:2006.

[197] Lawrence E. Nielsen, *J Appl Poly Sci*, 10(1); 97–103: 2003.

[198] Manuchehr Nikazar, Babak Safari, Babak Bonakdarpour, and Zohreh Milani, *Indian Polym J*, 14(12)1050–1057, 2005.

[199] Yoshito Ohtake, Tomoko Kobayashi, Hitoshi Asabe, Nobunao Murakami, Katsumichi Ono, *J Appl Polym Sci*, 70: 1643–1659, 1998.

[200] Otey F.H., Westhoff R.P., Doane W.M., *Indus & Eng. Chem.*, 26(8); 1659–1663:1987.

[201] J. Aburto, S. Thiebaud, I. Alric, E. Borredon, D. Bikiaris, J. Prinos and C. Panayiotou, *Carbo Polym*, 34(1-2); 101–112: 1997.

[202] D. Bikiaris, E. Pavlidou, J. Prinos, J. Aburto, I. Alric, E. Borredon and C. Panayiotou, *Polym Degrad Stab*, 60(2-3); 437–447:1998.

[203] Matzinos P, Tserki V, Gianikouris C, Pavlidou E & Panayiotou C, *Euro Polym J*, 38, 1713–1720:2002.

[204] Matzinos P, Tserki V, Kontoyinnis A, & Panayiotou C, *Polym Degrad Stab*, 77, 17–24: 2002.

[205] Soon-Do Yoon, Sung-Hyo Chough, Hye-Ryoung Park, *J Appl Poly Sci*, 100(5); 3733–3740: 2006.

[206] Soon-Do Yoon, Sung-Hyo Chough, Hye-Ryoung Park, *J Appl Polym Sci*, 100(3); 2554–2560:2006.

[207] Jun T. Kim, Dong S. Cha, Gee D. Lee, Tae W. Park, Dong K. Kwon, Hyun J. Park, *J Appl Poly Sci*, 83(2); 423–434: 2001.

[208] Jitendra K. Pandey, K. Raghunatha Reddy, A. Pratheep Kumar and R.P. Singh. *Polym. Degrad. Stab.*, 88(2); 234–250:2005.

[209] Jitendra K. Pandey, and R. P. Singh, *Biomacromo*, 2 (3), 880-885:2001.

[210] Plackett, David. Danish Polymer Centre, DTU, Lyngby, Den., Medical Plastics, 2002, Papers of the Annual Conference and Seminar, 16th, Copenhagen, Denmark, Aug., 27–30, 2002, 22–29. Hexagon Holding ApS, Copenhagen, Den CODEN: 69EPC2.

[211] S. Jerachaimongkol, V. Chonhenchob1, O. Naivikul and N. Poovarodom, *Kasetsart J. (Nat. Sci.)*, 40(5); 148–151:2006.

[212] L Contat-Rodrigp, A. Ribes Greus, *J Appl Polym Sci*, 83; 1683–1691:2002.

[213] D. Raghavan and A. Emekalam, *Polym Degrad Stab* 72(3); 509–517:2001.

[214] R. T. Nagaralli, A. Ramakrishna & A. Varada Rajulu, *J Plastic Film Sheet*, 13(3); 252–262:1997.

[215] Jo Ann Ratto, Peter J. Stenhouse, Margaret Auerbach, John Mitchell and Richard Farrell, *Polymer*, 40(24); 6777–6788:1999.

4.1 References

[216] X. Ren, *J Clean Produ*, 11; 27–40:2003.

[217] C. Rivard, L. Moens, K. Roberts, J. Brigham, and S. Kelley, *Enzy. Micro. Techno.* 17; 848–852:1995.

[218] A.G. Pedroso and D.S. Rosa, *Carbo Polym*, 59(1); 1–9:2005.

[219] Daniela Schlemmer, Maria J.A. Sales and Inês S. Resck, *Carbo Polym*, 75(1); 58–62:2009.

[220] Piyush B. Shah, S. Bandopadhyay and Jayesh R. Bellare. *Polym. Degrad. Stab.*, 47(2); 165–173:1995.

[221] Krishna Sastry, D. Satyanarayana, D.V.M. Rao, *J Appl Polym Sci*, 70; 2251–2257,1998

[222] R.R.N. Sailaja and S. Seetharamu, *React Func Polym*, 68(4); 831–841:2007.

[223] R. R. N. Sailaja, Manas Chanda, *J Appl Polym Sci*, 86: 3126–3134, 2002.

[224] B. G. Girija, R. R. N. Sailaja, *J Appl Polym Sci* 101; 1109–1120: 2006.

[225] Siddaramaiah, K. H. Guruprasad, R. T. Nagaralli, H. Somashekarappa, T. N. Guru Row, *J Appl Polym Sci*, 100; 2781–2789:2006.

[226] Baldev Raj, V. Annadurai, R. Somashekar, Madan Raj and S. Siddaramaiah, *Euro Polym J*, 37(1); 943–948:2001.

[227] Supong Arunvisut, Sutthipat Phummanee, Anongnat Somwangthanaroj, *J Appl Polym Sci*, 106; 2210–2217:2007.

[228] Bong Gyu Kang, Sung Hwa Yoon, Suck Hyun Lee, Jae Eue Yie, Byung Seon Yoon, Moon Ho Suh, *J Appl Polym Sci*, 60(11); 1977–1984:1998.

[229] Emo Chiellini, A Corti and G, *Polym Degrad Stab*, 81(2); 341–351:2003.

[230] R.N. Tharanathan, *Trends Food Sci Techno*, 14; 71–78:2003.

[231] Viktória Vargha and Patricia Truter, *Eur o Polym Jour*, 4(4); 715–726:2005.

[232] United States **Patent Application** 20080287592.

[233] K.Seethamraju, M.Bhattacharya, U.R Vaidya and R. G Fulcher, *Rheol Acta*, 33; 553:1994.

[234] U.R. Vaidya and M.Bhattacharya, *J Appl Polym Sci*, 52; 617:1994.

[235] M. Bhattacharya, U.R. Vaidya, D Zhang and R Narayan, *J Appl Polym Sci*, 57; 539:1995.

[236] U. R. Vaidya, M Bhattacharya and D. Zhang, *Polymer*, 36; 1179:1995.

[237] Z.H. Yang and M.Bhattacharya and U. R. Vaidya, *Polymer*, 37; 2137:1996.

[238] IdaVašková, Pavol Alexy, Peter Bugaj, Anna Nahálková, Jozef Feranc, Tomáš Mlynský, *Acta Chimica Slovaca*, 1(1); 301–308: 2008.

[239] N. St. Pierre, B. D. Favis, B. A. Ramsay, J. A. Ramsay and H. Verhoogt, *Polym*, 38(3); 647–655:1997.

[240] S. M. de Lima, M. A. G. A. Lima, S. M. Sarmento, G. M. Vinhas, Y. M. B. de Almeida, *Enpromer*, 2005.

[241] Glória Maria Vinhas; Suzana Moreira de Lima; Lívia Almeida Santos; *Braz Arch Biol Technol*, 50(3);Curitiba 2007.

[242] J. J. G. van Soest, K. Benes, D. de Wit and J. F. G. Vliegenthart, *Polymer*, 37(16); 3543–3552:1996.

[243] J. J. G. van Soest, S. H. D. Hulleman, D. de Wit and J. F. G. Vliegenthart, *Carb. Polym.,* 29(3), 225–232:1996.

[244] Jeroen J. G. van Soest, S. H. D. Hulleman, D. de Wit and J. F. G. Vliegenthart, *Indu Crops Prod,* 5(1); 11–22:1996.

[245] Parvinder S. Walia, John W. Lawton, Randal L. Shogren, *J Appl Polym Sci* 84: 121–131, 2002.

[246] Xiu-Li Wang, Ke-Ke Yang, and Yu – Zhong Wang, *J Macromol Sci*, C43(3); 385–409:2003.

[247] W. Liu, Y. J. Wang, Z. Sun. *J Appl Poly Sci,* 88; 2904–2911:2003.

[248] H. M. Wilhelm, M. R. Sierakowski, G. P. Souza and F. Wypych, *Carb Polym*, 52(2); 101–110:2003.

[249] S. St. Lawrence, P. S. Walia, F. Felker, J. L. Willett, *Polym. Eng. Sci.*, 44; 1839–1847: 2004.

[250] Williams PA, Ahmed FB., Douvlier JL, Durand S, Buleon A., *Carbo Polym* 38; 361–370:1999.

[251] Williams PA., Williams PA., *J Agri Food Chem*, 46; 4060–4065:1998.

[252] Martina Wollerdorfer and Herbert Bader, *Indust Cr ops Pr od,* 8(2); 105–112:1998.

[253] S. M. Goheen, R. P. Wool, *J. Appl. Polym. Sci42(10); 2691–2701:*,2003.

[254] R. Santayanon, J. Wootthikanokkhan, *Carbo Polym*, 51; 17–24:2003.

[255] Chin-San Wu, *Polym Degrad Stab*, 80(1); 127–134:2003.

[256] Xiaodong Zhang, Wenying Liuxin Li, *J Appl. Polym. Sci .*, 88; 1563–1566:2003.

[257] HanGuo Xiong, ShangWen Tang, HuaLi Tang and Peng Zou, *Carbo Polym*, 71(2); 263–268:2008.

[258] Lee Wook Jang, Eung Soo Kim, Hun Sik Kim, Jin-San Yoon, *J Appl Polym Sci*, 98; 1229–1234:2005.

[259] Maolin Zhai, Fumio Yoshii and Tamikazu Kume, *Carbo Polym*, 52(3), 311–317:2003.

[260] Zainuddin, Mirzan Thabran Razzak, Fumio Yoshii, Keizo Makuchi, *J Appl Polym Sci*, 1999; 1283-1290:1999.

[261] Xiudong You, Li Li, Jianping Gao, Jiugao Yu, Zhuxyan Zhao, *J Appl Polym Sci*, 88; 627–635: 2003.

[262] Zobel, HF, *Starch*, 40; 44, 1988.

5

Characterization Techniques

INSIDE THIS CHAPTER

5.0 Characterization Techniques

5.1 Physico-chemical Analysis

    5.1.1 FTIR: ATR Fourier Transform Infrared: Attenuated Total Reflectance: Spectroscopy

5.2 Thermal Analysis

    5.2.1 DSC: Differential Scanning Calorimetery

    5.2.2 TGA: Thermo Gravimetric Analysis

        5.2.2.1 Application of TGA

    5.2.3 Thermo Mechancial Analyser (TMA)

    5.2.4 Dynamic Mechanical Analyser (DMA)

5.3 Crystallographic Study

    5.3.1 XRD: X-Ray Diffraction

5.4 Morphology Analysis

    5.4.1 SEM: Scanning Electron Microscopy

Contd...

5.5 Water Absorption Analysis

5.6 Physico-mechanical Properties

    5.6.1 Tensile Strength

    5.6.2 Tear Strength

    5.6.3 Burst Strength

5.7 Degradation Study

    5.7.1 Weight Loss Study

    5.7.2 Carbonyl Index Study

    5.7.3 Mechanical Properties After Degradation

    5.7.4 Morphology Study After Degradation

5.8 References

5.0 CHARACTERIZATION TECHNIQUES

Biodegradable polymers are extensively used in different fields for many applications, for example as biodegradable sutures, surgical implants, agricultural mulches, packaging film, agrochemicals, products for one time use applications. The properties of biodegradable polymers to be used for different applications are determined by appropriate analysis of samples.

The analysis has to be done to assess the properties of the biodegradable polymer developed and the degree of environmental biodegradation expected after their life time. The assessment of samples may be done through various characterization techniques, which can be used for the study of identification of the functional group, structural properties, thermal behaviour, crystallographic nature, mechanical properties, water absorption properties and morphological properties etc.

Different types of analysis applied for the determination of various properties of biodegradable polymer samples are detailed here alongwith some examples of instruments used, their measuring principle, sampling procedures and ASTM methods applicable etc.:

5.1 PHYSICO-CHEMICAL ANALYSIS

Physico-chemical properties of samples can be analyzed though FTIR:ATR technique.

5.1.1 FTIR: ATR Fourier Transform Infrared: Attenuated Total Reflectance: Spectroscopy [1, 2]

Infra Red (IR) Spectroscopy is used for the determination of molecular structural entities in the biodegradable polymers.

Generally ASTM D5576-94 method is used for the determination of structural entities in polyolefins by Fourier Transform Infrared Spectroscopy. FTIR:ATR reveals the structure of a new compound and predict the presence of certain functional groups, absorbed at definite frequencies. The shift in the absorption position helps in predicting the factors, which are responsible for the shift. This method is applicable for samples those are too opaque or too thick for standard transmission methods. The technique is rapid, simple and requires a little sample preparation. Also, one of the major advantages of the ATR technique is that the spectrum obtained is independent of the sample thickness. Typically, the reflected radiation penetrates the sample to a depth of only few microns. Consequently, the method is useful for surface analysis. ATR spectra of compounds are plotted as % transmittance (%T) against wave number (υ). FTIR: ATR analysis can be performed over degraded samples to measure the extent of bio-degradation.

Principle

Figure 5.1 depict the FTIR instrument and ATR Accessory. An attenuated total reflection accessory operates by measuring the changes that occur in a totally internally reflected infrared beam, when the beam comes into contact with a sample as indicated in Fig. 5.3 an infrared beam is directed onto an optically dense crystal with a high refractive index at a certain angle.

This internal reflectance creates an evanescent wave that extends beyond the surface of the crystal into the sample held in contact with the crystal. It can be easier to think of this evanescent wave as a bubble of infrared that sits on the surface of the crystal. This evanescent wave protrudes only a few microns (0.5 μ – 5 μ) beyond the crystal surface and into the sample. Consequently, there must be good contact between the sample and the crystal surface. In regions of the infrared spectrum where the sample absorbs energy, the evanescent wave will be attenuated or altered. The attenuated energy from each evanescent wave is passed back to the IR beam, which then exits the opposite end of the crystal and is passed to the detector in the IR spectrometer. The system then generates an infrared spectrum.

FTIR Instrument

ATR Accessory

Internal Sketch of FTIR

Fig. 5.1 FTIR Instrument and ATR Accessory.

For the technique to be successful, the following two requirements must be met:

- The sample must be in direct contact with the ATR crystal, because the **evanescent wave** or bubble only extends beyond the crystal **0.5 µ – 5 µ**. The ATR accessories have devices that clamp the sample to the crystal surface and apply pressure.

- The refractive index of the crystal must be significantly greater than that of the sample. Typically, ATR crystals have refractive index values between 2.38 and 4.01 at 2000 cm^{-1}. Majority of solids and liquids have much lower refractive indices.

Figure 5.2 shows schematically the path of ray of light for total internal reflection. The ray penetrates a fraction of a wavelength (d_p) beyond the reflecting surface into the rarer medium of reflective index

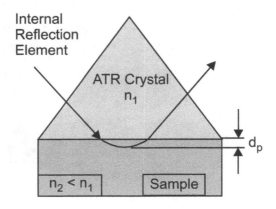

Fig. 5.2 Internal Diagram of FTIR:ATR.

5.1 Physico-chemical Analysis

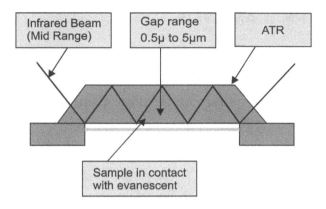

Fig. 5.3 Attenuated Total Internal Reflectance (ATR).

n_2 and there is a certain displacement (D) upon reflection. The Fig. 5.3 shows attenuated total internal reflectance over the sample in contact with gap range from 0.5 to 5µm size.

Instruments:

Models – **Nicolet 380 Spectrometer** (Thermo Scientific)

– OMNIC spectroscopy software

– **Spectrometer** (Perkin Elmer)

Sample:

Since the sampling is independent of sample thickness, samples of biodegradable polymers can be analysed using standard sample preparation procedure.

Procedure:

The equipment should be set to operate in the range of 600 to 4000 cm^{-1}. Infrared spectra has to be obtained in transmission mode with a Nicolet FTIR:ATR.

5.2 THERMAL ANALYSIS

Thermal analysis of polymers provides information about the thermal stability of compositions. Thermal analysis of biodegradable polymer can be done using Differential Scanning Calorimetery (DSC), Thermal Graviematric Analysis (TGA), Dynamic Mechanical Analysis (DMA) and Thermo Mechanical Analysis (TMA).

Following parameters of any biodegradable polymer can be studied by Thermal analysis:

(*a*) Melting Point using DSC

(*b*) Glass transition temperature using DSC, DMA and TMA.

(*c*) Degree of Cure using DSC.

(*d*) Decomposition temperature using DSC and TGA.

(*e*) Component quantification using TGA

(*f*) Percentage Crystallization using DSC

(*g*) Co-efficient of thermal expansion using DSC.

5.2.1 DSC: Differential Scanning Calorimetery

Differential Scanning Calorimetery (DSC) is the most widely used technique for analytical investigation of polymeric materials. It measures the differences in energy inputs into a substance and a reference material as they are subjected to a controlled temperature program, because practically all physical and chemical processes involves changes in enthalpy or specific heat [2]. In other words all materials have a finite heat capacity, heating or cooling a sample results in a flow of heat in or out of the sample.

5.2 Thermal Analysis

Figures 5.4 and 5.5 depict the DSC Instrument and its block diagram. Commercial DSC are of two types: heat flux DSC and Power Compensation. Heat flux DSC deploys common heating for sample and Reference. The differential heat flow to sample is proportional to the temperature difference that develops between Sample and Reference junctions of a thermocouple. However, in case of Power compensation temperature is controlled around Sample and Reference individually. Differential energy flow necessary to maintain both Sample and Reference on specified temperature program is recorded.

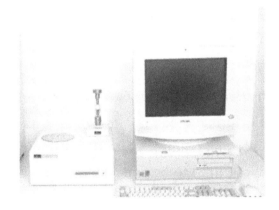

Fig. 5.4 DSC Instrument Perkin Elmer make.

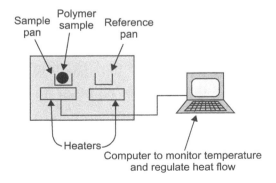

Fig. 5.5 DSC Instrument's block diagram.

Instruments:

Models – **Diamond, DSC** (Perkin Elmer)

– Pyris Kinetic Software

– **DSC 7** (Perkin Elmer)

- Sample and Reference pans are heated by two independent furnaces embedded in a temperature-controlled heat sink.
- DSC Measures the difference in Energy Inputs into a substance and a reference material as they are subjected to control temperature program.
- GEOMETERY and SIZE of Sample and Reference Cells are important because the Heat Flow depends on geometry and Thermal Diffusivities.

Sample Preparation for DSC Measurement: Samples could be taken in the form of powder or granule, molded or pelleted samples or film or sheet samples. 5 to 10 mg samples, to an accuracy of 0.01 mg are required to be encapsulated in hermetically sealed aluminium pan and purging of Nitrogen gas is done at a rate of 30 ml/min.

DSC Procedure: Standard procedure to perform DSC.

1. THREE RUNS are performed for each sample analysis.
2. 1st Heat sample at a rate of 10°C min^{-1} from ambient to 300°C above the melting point or to temperatures high enough to remove previous thermal history.
3. Hold temperature for 10 minutes.
4. Cool to at least 50°C below the peak crystallization temperature at a rate of 10°C min^{-1}.

5.2 Thermal Analysis

5. Repeat the process as 2nd Heating cycle at the rate of 10°C min⁻¹.

6. Scan thermal transition from temperature 1 to temperature 2 and in the subsequent cycles performed.

7. Apply same temperature profile to all samples.

8. Perform process under Nitrogen atmosphere.

9. Calculate heat of fusion and heat of crystallization using equation (5.1).

10. Select appropriate X and Y axis sensitivities to yield an area of 30 to 60 cm² (5 to 10 in².) under the fusion endotherm.

$$H = \frac{ABT}{W} * \frac{H_s W_s}{A_s B_s T_s} \qquad \ldots (5.1)$$

where:

H = Heat of fusion or crystallization of sample, J/kg (mcal/mg)

H_s = Heat of fusion or crystallization of standard J/kg (mcal/mg)

A = peak area of sample cm² (in.²)

A_s = peak are of sample cm² (in.²)

W = weight of sample, mg,

W_s = weight of standard, mg,

T = Y-axis sensitivity of sample, mW/cm (mcal/s per in. of chart)

T_s = Y-axis sensitivity of standard, mW/cm (mcal/s per in. of chart)

B = X-axis sensitivity (time base) of the sample, min/cm (min/in.)

B_s = X-axis sensitivity (time base) of the standard, min/cm (min/in.).

5.2.1.1 Applications of DSC

- Melting study
- Crystallization study
- Glass transition temperature study
- Crystallinity study
- Oxidative stability study
- Blend analysis
- Curing study.

5.2.1.1.1 Melting Study

The determination of melting point of polymers is required to find out the processing temperature of the semi crystalline polymeric material. It is a tool to distinguish different materials in a polymeric group. Under the influence of heat, most semi crystalline polymer

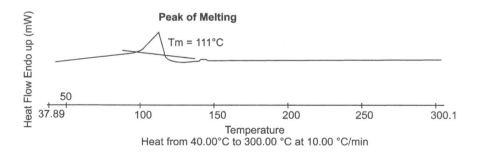

Fig. 5.6 DSC curve of polymer.

samples begin to melt. Basically melting destroys the crystalline nature of polymers. Polymer melt over a range of temperature in which onset of melting endotherm is often quoted as melting temperature. The Fig. 5.6 depicts the DSC curve of polymer.

5.2.1.1.2 Crystallization Study

Semi crystalline polymers when cooled at different programmed rate from melt will result in polymers with different degree of crystallinity. Slow rate of cooling will result into thicker lamella formation of crystalline fraction. Higher rate of cooling results in formation of thinner lamella formation with non crystalline fraction. Samples with large lamella thickness have higher onset melting temperature and heat of melting.

Lamellae thickness depends upon the rate of cooling. Polymers having large lamellae thickness have higher onset melting temperature as well as melting heat. However, rapid cooling produces crystal morphology with thinner lamellae. At the other end, slow cooling encourages the crystal morphology to form thicker lamellae.

5.2.1.1.3 Glass Transition Temperature Study

Glass transition temperature is the temperature at which the polymer sample transform from a glassy state to a rubber state. It is symbolically represented as T_g.

$$\text{Glassy state polymer} = \frac{\text{Temperature}}{T_g} \text{ Rubber state polymer.}$$

The transformation is accompanied by changes in the physical properties of the material. At T_g polymer undergoes enthalpies of relaxation, which appear as a positive deviation from DSC baseline resulting from the increase in specific heat above T_g.

The glass transition is a property of only the amorphous portion of a semi-crystalline solid. The crystalline portion remains crystalline during the glass transition. In glassy state the molecules in amorphous region are frozen on place. They may vibrate but can not have segmental motion in which they can move. Below glass transition temperature amorphous region of polymer remains hard, rigid and brittle.

If the polymer is heated it eventually will reach its glass transition temperature. At this temperature portions of the molecules can start to wiggle around. The polymer now is in its rubbery state. The rubbery state lends softness and flexibility to a polymer.

A given polymer sample does not have a unique value of T_g because the glass phase is not at equilibrium. The measured value of T_g will depend on:

(*a*) Molecular weight of the polymer,

(*b*) Thermal history

(*c*) Age

(*d*) Measurement method

(*e*) Rate of heating or cooling

5.2.1.1.4 Crystallinity Study

The change of enthalpy is usually a linear function of the reaction co-ordinate. A DSC endothermic or exothermic peak is shown by upward and downward curves and a DSC measurement gives the rate of change of enthalpy so that the area between a DSC curve and its extrapolated baseline indicates the total heat of reaction.

The crystallinity X_c is calculated by relative ratio of the enthalpy of fusion per gram of samples (ΔH_m) to the heat of fusion (ΔH_m^+)

5.2 Thermal Analysis

which is the heat of melting of the theoretical 100% crystalline polymer according to the equation 5.2 [3].

% Crystallinity $X_c = (\Delta H_m / \Delta H_m^+) \times 100$ ... (5.2)

ΔH_m : Area of endothermic melting peak directly equal to the heat of melting in Joule/gm.

ΔH_m^+ : Heat of melting (Theoretical) for virgin crystalline sample in Joule/gm.

5.2.1.1.5 Blend Analysis

DSC can be used to study the blend analysis quantifying and identifying the number of components present. The heat of melting or melting endothermic peak can identify blending polymers. These peaks indicate the amount present of each component in the blend.

5.2.1.1.6 Curing Study

On heating thermoset material above T_g cross linking reaction is initiated in which the growth and formation of molecular due to branching and cross linking between the molecules take place. Which leads to formation of completely crosslinked network. The reaction is highly exothermic.

Degree of Cure (%) = $(H_c / \Delta H_c^+) \times 100$ ... (5.3)

ΔH_c : Measured heat of curing of samples in Joule/gm.

ΔH_c^+ : Heat of curing of a fully cured sample in Joule/gm.

5.2.2 TGA: Thermo Gravimetric Analysis

The thermal study provides the information about the thermal and oxidative stability of the polymers. TGA is used to measure mass flow into a sample and out of sample. It measures the change in the mass of a sample as it is heated, cooled or held isothermally (at a constant temperature). Therefore TGA can detect percentage of moisture and volatile percentage of fillers and additives in the multi component system, compositional difference in blends, degradation temperature and decomposition kinetic study of a sample. During this study sample weight is continuously monitored with increase in temperature either at constant rate or through a series of steps. Components of polymers decompose at different temperature. This leads to a series of weight - loss steps that allows the components to be quantitatively measured.

Analyser for TGA is often referred to as a thermo balance. There are two essential components:

(*i*) Furnace system

(*ii*) Balance system.

It informs about the thermal degradation, kinetics of materials and service lifetime prediction along with the thermal breakdown as a function of time in an instrument called thermo-balance.

The analytical results are a plot of the mass or the percentage of original mass remaining at specific temperature or time. The weight losses accompanying the de-volatilization and decomposition process in TGA do not occur at precisely defined transition temperatures but occur in ranges that are highly dependent on both the physical and chemical characteristics of the material and as well as on instrumental and environmental factors. Weight changes are measured using a highly sensitive deflection type

5.2 Thermal Analysis

thermo-balance capable of reproducing mass changes as small as ± 1.0 µg and depending upon the objective of the experiment, the sample can be bathed in an inert atmosphere using nitrogen or argon or it may be statically or dynamically contacting with an oxidizing or reducing gas. TGA in an inert atmosphere is used as a method for empirically assessing the thermal stabilities of experimental polymers.

The shape of the thermo-gram gives the clue about the nature of decomposition. TGA describe the thermal stability of the polymer.

Instrument

Model - Perkin Elmer Thermal Analyser

Fig. 5.7 Perkin Elmer's Thermo-gravimetric Analyzer.

TGA Procedure

Thermal degradation of samples is analysed in the dynamic mode and weight loss of samples is subject to controlled temperature program. The samples are taken in a crucible situated in the electric furnace and heated from room temperature 40–500°C. The approximate 6–10 mg of each sample is used for analysis. Each sample is subjected to heating rate of 10°C/min under nitrogen atmosphere. Prior to conduct the experiment approximately 10 milligram samples are dried in vacuum oven at 60°C for 2 hours. The onset of degradation temperature is recorded for each sample.

5.2.2.1 Application of TGA

- Determination of mass loss in composites/blends
- Thermal stability assessment
- Percentage weight loss of components
- Percentage of moisture and volatiles
- Percentage of fillers and additives
- Degradation temperature and decomposition kinetics study.

5.2.2.1.1 Derivative Thermograviemetry (DTG) Study

It is defined as the first derivative of the mass change of the sample as a function of temperature and time with respect to time. Area under the DTG curve is directly proportional to the mass change. Figure 5.8 shows DTG curve between rate of mass loss versus temperature. The DTG ordinate has unit of mass per unit time (g/min). These DTG curves represent mass loss and are very accurate in determining the unambiguous determination of each component.

5.2 Thermal Analysis

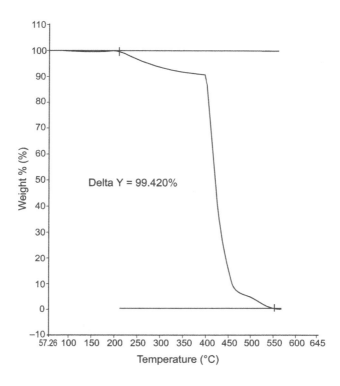

Fig. 5.8 DTG Curve between rate of Mass Loss and Temperature.

5.2.3 Thermo Mechancial Analyser (TMA)

The technique is used to monitor the dimension of a substance under negligible load conditions as a function of temperature and time with controlled pressure conditions. The temperature program involves controlled heating or cooling or maintaining a constant temperature usually in an inert atmosphere.

Thermo Mechanical Analyser has four important components.

(a) Linear static force monitor to apply static load to ensure probe contact with sample.

(b) Linear variable displacement transducer (LVDT) to monitor the change in the dimension of the sample.

(c) Probe and Platform to mount the sample

(d) Micro furnace to heat and cool the sample

Slightest change in the dimension is recorded.

Applications of TMA

It is useful means to detect following:

1. Coefficient of thermal expansion
2. Contraction
3. Shrinkage
4. Foaming and softening points of material.

5.2.4 Dynamic Mechanical Analyser (DMA)

Principle

The technique in which phase angle and amplitude of a uniform sample is analysed under a selected dynamic and static load as a function of temperature or time subject to controlled temperature program in a controlled atmosphere.

Phase angle: Phase angle is the time it takes for the material to respond to the applied stress. Units of phase angle are degree or radian, since it is based on Sine Wave.

Amplitude: Amplitude is the magnitude of dynamic displacement of the sample resulting from the applied dynamic load. It is measured from the means to the extreme position of the oscillatory movement.

Phase lag: Lag is the angular difference between the applied dynamic load and measured strength of the material response.

5.2 Thermal Analysis

Proper mechanism to introduce phase lag and angle is to consider three types of material properties: perfectly elastic, perfectly newtonian fluid and visco elastic.

DMA Analyser

This is a methodology where the phase angle (the time lag between the applied dynamic stress and measured strain by the system) and amplitude (dynamic displacement occurs in the system) of a uniform sample geometry is monitored under a selected dynamic and static load as a function of time or temperature under strict control of temperature program under controlled atmospheric conditions. Temperature could be heating or cooling or constant either in oxidizing or inert atmosphere. Modes are selectable as dynamic or tension or compression as per the requirement.

DMA applies user selectable dynamic load on to sample through electronic probes and slightest change of dynamic displacement and phase lag is recorded. This is done through position sensors under controlled conditions.

Important components of DMA Analyser are:

1. Force motor to apply
2. The static and dynamic load
3. Linear variable displacement transducer (LVDT) monitoring change in dimension and phase angle.
4. Measuring system and Micro furnace

Operational behaviour of DMA

Perfectly Elastic Material: A perfectly elastic material when subjected to a dynamic load, it exhibits no phase lag between the applied dynamic load and the measured strain response. Here, the point of maximum strain measured coincides with the maxima of the applied dynamic load. Spring behaviour used to model it perfectly. On application of load, the material will be displaced by an amount of deflection. On removal of load, the material will return back to its original position. Indirectly, the energy applied is completely recovered.

Perfectly Newtonain Fluid: It is an example of perfectly inelastic material subject to dynamic load. The measured strain response lags behind the applied dynamic load by a phase angle of 90°. A dashpot is used to model perfectly inelastic material. On application of a load, the material gets displaced by the same amount of deflection. On removal of load, the material will not return back to its original position and it can be observed that instead of some recovery of energy, the entire energy has been lost in the process.

Viscoelastic Material: In such materials the measured strain lags behind the applied dynamic load. The material called visco elastic are those that have a phase angle between 0 to 90°. Here, on application and removal of load partial displacement and partial return take place. Confirming that energy supplied to the material is partially lost.

The DMA can provide the following properties of materials:

(*a*) Storage modulus

(*b*) Loss modulus

(*c*) Tangent delta

(*d*) Complex modulus

5.2 Thermal Analysis

(e) Storage compliance

(f) Loss compliance

(g) Complex compliance

(h) Shear storage modulus

(i) Shear loss modulus

(j) Shear complex modulus

(k) Storage viscosity

(l) Loss viscosity

(m) Complex viscosity

Applications of DMA

1. Characterization of Cured Epoxy Samples

The degree of cure of epoxies measured by glass transition temperature measurement through DMA, TMA or DSC. The peak of tangent delta plot as a function of temperature provides us less ambiguous glass transition temperature. DMA is more sensitive to low energy sub T_g transitions

2. Gel point of PVC

Polymer Network is to form cross linking linear chains. The application of heat initiates this process and would result molecules of infinite molecular weight. The point at which this infinite network is formed called as gel point. The gel point is detectable through DMA at a point where tangent delta = 1.

3. Characterization of Epoxidized Natural Rubber

The storage modulus of natural rubber epoxidixed at different molecular percentage can be easily detected when subjected to constant dynamic stress oscillating at a linearly changing frequency at ambient conditions.

(*i*) Stress strain relationship of materials: Study of stress and strain relationship can be done by DMA for wide range of materials with some special features.

(*ii*) Creep recovery of polymer: Creep is time dependent strain monitored, when a visco elastic material is subjected to a constant load at a defined temperature. Relaxation of strain is called recovery when load is partially or completely removed. When there is sudden constant stress on the sample, sample experiences an instantaneous deformation, followed by primary and secondary creep. On removal of stress sample recovers partially.

5.3 CRYSTALLOGRAPHIC STUDY
5.3.1 XRD: X-Ray Diffraction

Crystallization behaviour of biodegradable compositions can be studied by Differential Scanning Calorimeter and *X*-Ray Diffraction. *X*-rays are produced by the bombardment of anodes with high energy electrons. The laboratory equipment employed with *X*-ray scattering generally involves an *X*-ray tube with CuK_α ($\lambda = 0.154$ nm). Nickel foil is employed to filter out $K\beta$ radiation. Scintillation detectors allow for the conversion of *X*-ray radiation into an electrical charge with a photomultiplier. Wide Angle *X*-ray Diffraction (WAXD) (10 to 60°) can yield structure determination less than several nanometers. X-ray incidence angle is θ.

WAXD of crystalline parameters exhibits sharp concentric rings indicating high degree of order. Amorphous polymers exhibit a diffuse halo, indicating lack of order. It is also employed to study

5.3 Crystallographic Study

the miscibility in polymer mixes, phase separated combinations results of molecular chain spacing. Miscible combination exhibits a larger molecular inter-chain spacing. Indicating a unique amorphous structure developed from the components different than either of the constituents. It uses Bragg's X-ray diffraction spectrometer following Bragg's law where $n\lambda = 2d \sin \theta$.

If n = 1, it is first order reflection (hkl = 111). It tells us at what angles X-rays will be diffracted by a crystal when the X-ray wavelength and distance between the crystal atoms are known.

Instrument:

Model – Philips – PW3710 (Fig. 5.9)

– PC APD Software

XRD Procedure

Radial continuous scans of intensity (I) versus diffraction angle (2θ) are recorded in the range of 10 to 60° at the rate of 2° per minute at room temperature. The generator is operated at 50 kV and 60 mA.

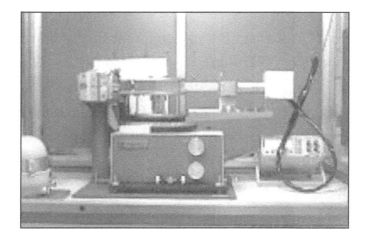

Fig. 5.9 X-ray Diffractometer.

Parameters Analysed

- Degree of Crystallinity
- Crystal Size
- d -spacings.

It determines the degree of Crystallinity in polymers from the relationship of the peak area to the total area using equation (5.4).

$$\text{Degree of Crystallinity } \alpha_x = \left[\frac{I_{cr}}{(I_{cr} + I_{am})}\right] \qquad \ldots (5.4)$$

Where I_{cr} is an integral intensity of the crystalline phase and I_{am} is integral intensity of the amorphous phase. The apparent dimensions of a crystallite D_{hkl} along the direction perpendicular to the crystal plane hkl can be determined using Scherrer's equation at equation (5.5). Bragg's diffraction occurs when radiation, with wavelength comparable to atomic spacings, is scattered in a seculer fashion by the atoms of crystalline system, and undergoes constructive interference.

$$D_{hkl} = \left[\frac{K\lambda}{\beta(\cos\theta)}\right] \qquad \ldots (5.5)$$

Where, D_{hkl} – Crystallite size along the direction perpendicular to reflection plane (*hkl*)(nm), θ – Bragg angle, λ – Wavelength of X-ray used (0.1542 nm), β – Peak width of diffraction beam used (rad), K - Shape factor of crystalline being related to the shape of a crystalline and definition of β, when β is defined as the half – height width of diffraction peak, $K = 0.9$.

Strengths of X-RD

- Non-distructive, small amount of sample
- Relatively rapid

- Identification of compounds/phases – not just elements.
- Quantification of concentration of phases
- Classification of powders, but solids also possible.
- Inform about crystallinity, size/strain, crystallite size and orientation.

5.4 MORPHOLOGY ANALYSIS

5.4.1 SEM: Scanning Electron Microscopy

SEM is a scientific instrument, which uses beam of energetic electrons to examine objects on a very fine scale.

The scanning electron microscopy (SEM) technique provides adequate information about the surface morphology, phase domains, pinholes defects and patterns and also about other topological features of the composition.

- **Topography:** The surface features of an object or "how it looks", its texture; Direct relation between these features and materials properties.
- **Morphology:** The shape and size of the particles making up the object; Direct relation between these structures and materials properties.
- **Composition:** The elements and compounds that the object is composed of and the relative amounts of them; Direct relationship between composition and materials properties
- **Crystallographic Information:** How the atoms are arranged in the object; Direct relation between these arrangements and material properties.

The scanning electron microscopy (SEM) technique provides adequate information about the surface morphology, phase domains, pinholes defects and patterns and also about other topological features of the composition.

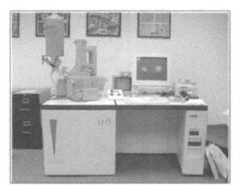 

Fig. 5.10 Scanning Electron Microscope Model No. LEO 435 VP.

SEM reveals the indirect information about the particle size, about the nature of cross linking between two polymers and also provides the information about the mixing pattern of the two polymers. SEM uses electrons to acquire an image. Electrons are directed towards the sample by a voltage bias. Focusing and magnification are carried out by magnetic lenses. By interactions within the sample, the electron beam is altered. These alterations are transformed to an image.

Principle (Principle of Functioning in Fig. 5.12)

A focused beam of electrons is rastered across a sample surface thereby generating secondary electron that are detected, which results in an image of the variation of secondary electron intensity with position on the sample. The variation is largely dependent on the angle of incidence of the focused beam onto the sample, thus yielding a topographical image with a high resolution (10 – 30 Å).

5.4 Morphology Analysis

This image is magnified by magnetic lenses. To avoid charging, insulated sample must be coated with a conductive film.

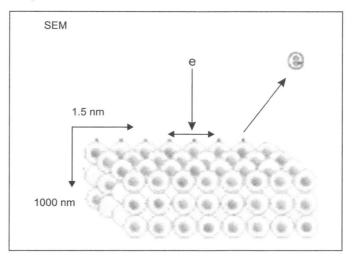

Fig. 5.11 Scanning Electron Microscope Image.

Instruments

    Model : **LEO 435 VP** (Fig. 5.10)

    Model : **JEOL 840** (Fig. 5.13)

Sample:

Small button type samples were used after fracturing for the surface analysis by SEM.

Procedure:

The Variable scanning electron microscope (SEM) model **LEO 435 VP/JEOL 840** can be used to examine the fractured surface of the samples. Each sample has to be washed with distilled water and dried in a vacuum oven at 60°C for 24 hours. Samples are fractured immediately after freezing in liquid nitrogen and are sputter coated with gold to prevent any electrical discharge during examination. The secondary electron image is used throughout using 15 kV accelerated potential. Micrographs of samples are taken at different magnifications to observe the micro-structural changes arising from the degradation phenomenon.

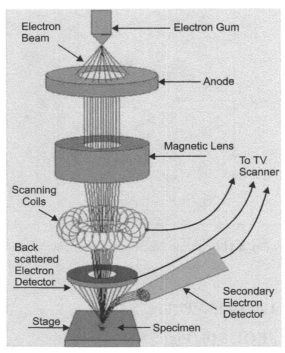

Fig. 5.12 Scanning Electron Microscope Principle.

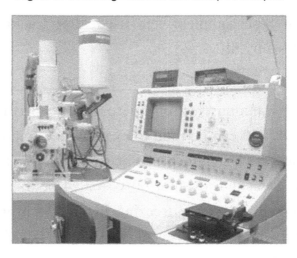

Fig. 5.13 Scanning Electron Microscope Model No. JEOL 840.

5.5 WATER ABSORPTION ANALYSIS

The water absorption of the samples is generally been measured as per ASTM D 570–63. The specimens in the sample of rectangular stripes *i.e.* 75 × 25 mm by size conditioned at 50±1 degree centigrade for 24 hours in an oven. Conditioned specimens are weighed (W_i) on a digital balance with a precision of 0.0001gram. The specimens are than immersed in distilled water at room temperature for 24 hours, removed, wiped dry by filter paper and weighed (W_w) immediately. After weighing the samples are again dried at 50±1 degree centigrade for 24 hours and finally weighed (W_f) again. The water absorption is calculated by equation (5.6). The average reading of three samples is required to be taken in both cases.

$$Water.Absorption,.\% = \left[\frac{W_w - W_f}{W_i}\right] \times 100 \qquad ...(5.6)$$

5.6 PHYSICO-MECHANICAL PROPERTIES

5.6.1 Tensile Strength

Mechanical properties such as tensile strength and elongation at break are important for all applications where the polymeric materials are used as structural materials and for packaging applications. The elongation at break is a useful index of ductility and hence the packaging durability. The tensile strength and elongation at break of the samples is determined using following methods:

ASTM D 638 for Sheets

ASTM D 882–67 for Films.

Parameters

Distance between grip for sample of films (thin samples) should be: 50 mm or more and thicker samples should be: *i.e.* 60 mm or more.

Speed of grip separation for thinner samples may be kept as: 500 mm/min and it may be kept as 50 mm/min for thicker samples. Width of the specimen should be: 10 mm. The most common type of stress strain measurement is by stretching the material. Nominal tensile stress can be expressed by the equation (5.7)

$$\text{Tensile Stress} = \frac{F}{A_o} \qquad \ldots (5.7)$$

where F is the tensile force applied at any given moment.

A_o is the area of the original cross section.

As the material stretches during tension its dimension orthogonal to axis of applied force decreases and so the area of cross section decreases. The tensile strength at break or the ultimate tensile strength is the maximum tensile stress sustained by the specimen at the moment of the rupture of the specimen. If the tensile strength induced a stretch to length l at the time of the failure of the test specimen the ultimate elongation or the elongation at the break is the strain produced in the test piece, expressed as the equation (5.8).

$$E_{break} = [\{l - l_o\}/l_o] \qquad \ldots (5.8)$$

where l is the length of test specimen at failure and

l_o is the original length of the specimen.

Test conditions for tensile test (an example of polymer specimen):

5.6 Physico-mechanical Properties

- The specimens conditioned for 24 hours at 60% RH and at room temperature 30°C.
- Tested in Universal Testing machine of 1600 kN capacity.
- Five samples analysed and strained at specific rate at room temperature 30°C.
- Average value of elongation at break observed.

5.6.2 Tear Strength

The tear strength is widely used as one of the indices of the tearing resistance of plastic films used in packaging applications. The tear strength has been measured using a tear strength tester.

The specimen I mounted in the jaws of the tear testing machine. The cut is initiated using a sharp blade provided in the machine. When the pendulum is released it swings down under the force of gravity under the specimen is torned from the already begin slit / notch, because of the energy required to tear the film. The pendulum has less energy then if it had fallen freely. The difference in energy is indicated by a pointer on a calibrated scale. The pendulum is released and the specimen is allowed to tear completely and reading is noted down to measure the tear strength. Tear strength is expressed in grams for thin samples. The Tear strength is measured in the machine direction and transverse direction.

5.6.3 Burst Strength

The strength of plastic is evaluated by means of simple bursting device. The burst strength of the thin sample is the resistance it offers to steadily increasing pressure applied at right angle to the surface under certain defined conditions. The burst strength is taken to be the pressure at the moment of failure, and is a measure of the capacity of the sample to absorb energy. Burst strength has been measured using – Burst Tester.

The specimens have been placed in between two angular rings of the machine. The inlet valve is opened to maintain an air pressure of 110 lbs/sq inch or so. The sample have been exposed to air pressure at a controlled rate and increased until it fails. The pressure at failure in lb/sq inch is taken as burst strength of the material.

5.7 DEGRADATION STUDY

Biodegradation **Assessment** of the prepared samples is done by conducting Weight loss analysis, Carbonyl Index analysis, Physico-mechanical properties analysis and Morphological analysis. The Physico-mechanical studies can not be conducted beyond a certain period because degradation level of samples becomes critical.

5.7.1 Weight Loss Study

Weight loss of the samples is determined by weighing the samples before and after biodegradation in compost. The percentage weight loss of samples is calculated using the equation (5.9).

$$Weight\ Loss,\ \% = \left[\frac{W_i - W_f}{W_i}\right] \times 100 \qquad \ldots (5.9)$$

where W_i is the initial weight of the sample before degradation and W_f is the final weight of the sample after degradation.

5.7.2 Carbonyl Index Study

The samples of various compositions are investigated for their degree of degradation through FTIR: ATR method. Specifically the Carbonyl Index of these samples is calculated and analyzed.

Sample:

Since the sampling is independent of sample thickness, samples of polymer/starch compositions can also be used for taking the FTIR: ATR of the degraded samples.

Procedure:

Infrared spectra obtained in transmission mode with a Nicolet make FTIR: ATR equipment. The equipment is set to operate in the range of 600 to 4000 cm^{-1}. Molecular degradation is characterized by carbonyl index (C.I.). It is computed as the relative areas under the carbonyl peak A_C (1700–1800 cm^{-1}) and a reference peak A_R not affected by exposure (centered at 1400–1500 cm^{-1}) by equation (5.10).

$$\text{C.I.} = \frac{A_c}{A_R} \qquad \ldots (5.10)$$

5.7.3 Mechanical Properties After Degradation

- **Physico-mechanical Properties:** Tensile test of rectangular film specimen (*i.e.* 10 × 180) mm can be conducted using Universal Testing Machine at a crosshead speed of say 50 mm/min according to ASTM D882. At least (05) five specimens of each sample should be tested.

5.7.4 Morphology Study After Degradation

- **Physical Appearance:** A scanning electron microscope for example model no. LEO 435 VP can be used to investigate the morphology of samples.

  Test Procedure:

- Sample should be washed with distilled water and dried in a vacuum oven at 60°C for 24 hours.

- The scanning electron microscope be operated at 15 kV.

- The film surface is sputter coated with gold prior to investigation to avoid surface charging under the electron beam.

5.8 REFERENCES

[1] Kemp, William. Organic Spectroscopy, third edition, New York, *Palgrave Publisher Limited.* p 72–75: 1991.

[2] ASTM D3417–99, standard test methods for enthalpies of fusion and crystallization of polymers of polymers by Differnetial Scanning Calorimetery, Vol. 8; Rubbers and Plastics.

[3] H.A. Khonakdar, S. H. Jafari, M. Taheri, U. Wagenkenecht, D. Jehnichen, L. Haussler, J. Appl Polym Sci., 100;3264-3271:2006.

ASTM Methods

1. ASTM D 4635-95 The standard specifications for polyethylene films made from low density polyethylene for general use and packaging applications.

5.8 References

2. ASTM D 570-63, Method used for the analysis of Water Absorption polymeric samples.

4. ASTM D774, Bursting Strength of Packaging Materials.

5. ASTM D 1922, Test Method for Propogation Tear Resistance of Plastic Fim and Thin Sheeting by Pendulum Method.

6. ASTM D 638 Standard Test Method for Tensile Properties of Plastics, 2001.

7. ASTM D 882 Standard Test Method for Tensile Properties of Thin Plastic Sheeting, 2001.

8. ASTM D 1238 for MFI Testing of Polymeric Materials.

9. ASTM D 1505 Test Method for Density for Density of Plastics by the Density-Gradient Technique Testing of Polymeric Materials.

10 ASTM D 3418–82 Standard Test Method for Transition Temperature of Polymers by Thermal Analysis.

❑❑❑

6

Biodegradation

> **INSIDE THIS CHAPTER**
>
> 6.0 Biodegradation
>
> 6.1 Degradation Process of Bio-degradable Polymer
>
> 6.2 Factors Affecting Biodegradability
>
> 6.3 International Standard Methods for Bio-degradability Testing
>
> 6.4 Methods of Biodegradation
>
>   6.4.1 Soil Burial
>
>   6.4.2 Composting
>
> 6.5 Definition of Composting
>
>   6.5.1 According to ASTM Standards
>
>   6.5.2 According to ISO Draft Standards
>
>   6.5.3 According to British Standards
>
>   6.5.4 According to US Environmental Protection Agency (US EPA)
>
> 6.6 Factors Affecting Composting Process
>
> 6.7 Phases of Composting Process
>
> 6.8 An International Standard related to Composting

Contd...

      6.8.1 ASTM Standards Related to Composting

      6.8.2 ISO Standards Related to Composting

      6.8.3 EN Standard Related to Composting

6.9 Methods of Biodegradability Testing by Composting

      6.9.1 ISO 14855–1 and 2

6.10 Methods of Biodegradation Testing

6.11 Biodegradation Supporting Environments

6.12 Duration of Bio-Degradation and Risk of their entry into Nature

6.13 Biodegradation Measurement Procedures

6.14 References

6.0 BIODEGRADATION

The terms 'biodegradation' and 'compostability' are very common but are frequently used. In biodegradation, the enzymes of the biosphere essentially take part at least in one step during cleavage of the chemical bonds of the material. Notably, biodegradation does not ensure that a biodegradable material will always degrade. In fact, degradation will only occur in a favourable environment, and the biodegradable material will not necessarily degrade within a short time. It is, therefore, important to couple the term biodegradable with the specification of the particular environment where the biodegradation is expected to happen, and of the time-scale of the process.

Some commodity plastics can be degraded at every stage of production in both the stabilized and unstabilized form and also undergo very slow biodegradation. It is now well evident that polymer/plastics waste management through biodegradation or bio-conversion is the most suitable solution for 'Plastic Waste Management' among the other traditional methods like Incineration, Pyrolysis, Land filling etc.

Several means [1-4], have been used to achieve the biodegradability/bio-deteriobility/Photo-biodegradability in polymers. There is a group of commercial biodegradable polymer representing combination of polymers from different origins. They have been formulated in such a way, so as to, offer interesting properties limiting the amount of costly materials in their composition. These materials are known as "environmentally acceptable degradable plastics" [5]. During their exploitation, these materials display a behaviour similar to that of conventional synthetic polymers. When their service time comes to an end, they are capable of degrading to low molecular weight compounds as a result of the combined action of different environmental agents- biological, chemical, and physical.

In agreement to above standards, the said degradation should lead to carbon dioxide, water, and cellular biomass in quantities similar to those of the natural biodegradable polymers, without the formation of permanent or toxic residues. Consequently the biodegradation definition must consider not only the degree of biodegradation, but also the impact of the polymer by-products on the environment.

Polysaccharides have been used more frequently as natural fillers for this purpose and well investigated by Griffin and Bastioli particularly in polyolefins at fairly low concentrations. The major problem is incompatibility between hydrophilic polysaccharides and generally the hydrophobic nature of polymer matrix where the complex mechanism of coupling through coupling agents, morphology of interfaces, surface energy and wetting phenomenon considerations are also necessary [6].

6.1 DEGRADATION PROCESS OF BIO-DEGRADABLE POLYMER

Biodegradable polymer generally degraded in two steps as primary and ultimate biodegradation. The biodegradation process is shown in Fig. 6.1.

During primary degradation, main chain cleavage occurs forming low molecular weight fragments that can be assimilated by microorganism. This molecular weight reduction is mainly caused by hydrolytic or main chain scission. The primary degradation use the environmental water with or without aid of an enzyme (biotic or abiotic conditions respectively). During secondary degradation the oligomers are incorporated into living cells for further assimilation:

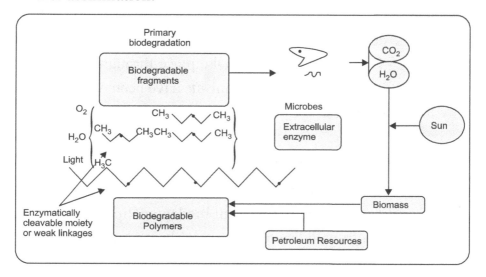

Fig. 6.1 Schematic Representation of the Degradation Process in Biodegradable Polymer [7].

- **Aerobic biodegradation:**

    In the presence of oxygen (under aerobic condition), this assimilation will produce carbon dioxide, water and the cells metabolic products.

$$C_{POLYMER} + O_2 \longrightarrow CO_2 + H_2O + C_{RESIDUE} + C_{BIOMASS} + SALTS$$

Anaerobic biodegradation:

In the absence of oxygen this assimilation would produce methane and others.

$$C_{POLYMER} \longrightarrow CO_2 + CH_4 + H_2O + C_{RESIDUE} + C_{BIOMASS} + SALTS$$

These assimilation processes, giving rise to inorganic materials, are called mineralization. Complete biodegradation occurs when no residue remains and complete mineralization occurs when original substrate $C_{POLYMER}$, is completely converted into gaseous products and salts [8]. In contrast to enzymatic degradation, where material is degraded gradually from surface inwards, chemical hydrolysis of a solid material can take place throughout its cross section. Different hydrolysis mechanisms have been reviewed, not only for backbone but for pendant groups too.

6.2 FACTORS AFFECTING BIODEGRADABILITY

- The accessibility of polymer to water-borne enzyme systems is vital because the first step in biodegradation involves the action of extra cellular enzymes which breakdown the polymer into products small enough to be assimilated. Therefore, the physical state of plastic and surface offered for attack are two important factors for bio-disintegration.

- Biodegradability is usually affected by the hydrophilic nature of polymer composition and the crystallinity of the polymer. A semi-crystalline nature of polymer composition tends to limit the accessibility, effectively confining the degradation to the amorphous regions of the polymer. However contradictory results, are reported. For example, highly crystalline starch materials and bacterial polyesters are rapidly hydrolyzed.

- Chemical linkages in the polymer compositions, linkages involving hetro atoms, such as ester and amide bonds are considered more susceptible to enzymic degradation.

- Pendant groups, their position and their chemical activity affect the biodegradability of polymer compositions.

- End groups and their chemical activity do contributes in biodegradation of polymer compositions.

- Molecular weight distribution of the polymer can have significant effect on rates on depolymerisation. This effect has been demonstrated for number of polymers, where a critical lower limit (minimum number of polymer units) must be present before the process starts.

- Interaction with other polymers, also affect the biodegradation process. These additional materials may act as a barrier to prevent migration of micro-organisms, enzymes, moisture or oxygen into the polymer domains of interest. It becomes a major barrier during initial stage of degradation.

- Susceptibility of a biodegradable polymer composition to microbial attack is sometimes decreased by grafting it onto a non-biodegradable polymer or by cross-linking. On the other hand, in literature it has been suggested that combining a non-biodegradable polymer with one that is biodegradable or grafting a biodegradable polymer onto a non-biodegradable polymer backbone polymer may result in a biodegradable system. Apparently, an environmentally degradable polymeric material is subjected to these factors simultaneously.

6.3 INTERNATIONAL STANDARD METHODS FOR BIO-DEGRADABILITY TESTING

There are several international standard methods for determination of Biodegradability of Biodegradable Plastics few of them are listed in Tables 6.1 and 6.2.

Table 6.1: ASTM Test Methods for Determining Biodegradability

1. ASTM D 5247 Determining the Aerobic Biodegradability of Degradable Plastics by Specific Microorganisms.
2. ASTM D 6002–96 Guide for Assessing the Compostability of Environmentally Degradable Plastics.
3. ASTM D 5338–98 Test Method for Determining Aerobic Biodegradation of Plastic materials under controlled composting conditions.
4. ASTM D 6340–98 Test Methods for Determining Aerobic Biodegradation of Radio labelled Plastic Materials in an Aqueous or Compost Environment.
5. ASTM D 5209 Test Methods for Determining the Aerobic Biodegradation of Plastic Materials in the presence of Municipal Sewage Sludge.
6. ASTM D 5210 Test Methods for Determining the Anaerobic Biodegradation of Plastic Materials in the presence of Municipal Sewage Sludge.
7. ASTM D 5152 Water Extraction of Residual Solids from Degraded Plastics for Toxicity Testing.

The International Organization for Standardization (ISO) is a worldwide federation of national standards bodies (ISO member bodies). Since the working group on biodegradability of plastics was created in 1993, rapid advances have been made in this area. Table 6.2 described the methods for aerobic biodegradation.

Table 6.2: ISO/DIS Methods for the Evaluation of Biodegradability.

Method No.	Description
ISO/DIS 14851	Evaluation of the ultimate aerobic biodegradability in an aqueous medium- method by determining the oxygen demand in a closed respirometer.
ISO/DIS 14852	Evaluation of the ultimate aerobic biodegradability in an aqueous medium-method by analysis of released carbon dioxide.
ISO/DIS 14855	Evaluation of the ultimate aerobic biodegradability and disintegration of plastics under controlled composting conditions-method by analysis of released carbon dioxide.

6.3 International Standard Methods for Bio-degradability Testing

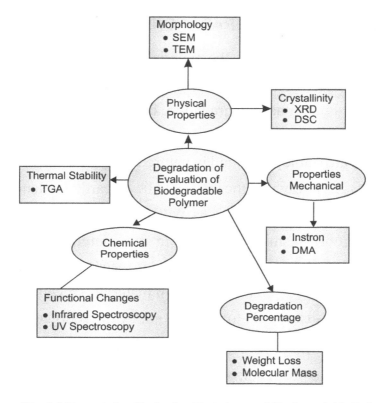

Fig. 6.2 Degradation Evaluation Techniques of Biodegradable Polymer.

These three ISO/DIS 14851, 14852 and 14855 are recognized as useful screening tests for establishing the aerobic biodegradability or compostability of plastics. Some others test methods are DIN (German) DIN 54900-Draft for Evaluation of the compostability and CEN (European) CEN TC 261/ SC4/ WG2 for Evaluation of the compostability, biodegradability and disintegration. Since all degradation processes decrease properties of polymers we can use almost all analytical techniques for the evaluation of degradability.

The characterization of bio-degradation can be carried out by several means those have been summarized in Fig. 6.2.

6.4 METHODS OF BIODEGRADATION

It is equally important as how the material is formed and how it is degraded. Generally the breakdown of polymer materials happens by microbial action, photo degradation, or chemical degradation. All three methods are classified under biodegradation, because the end products are stable and found in nature. Many biopolymers are designed in such a way that these have to be discarded in landfills, composts, or soil. These materials will be broken down, if the required microorganisms are present in the compost or landfill sites. Normal soil bacteria and water are generally required. Polymers having base of naturally grown materials (such as starch or flax fiber) are susceptible to degradation by microorganisms.

The material may or may not decompose rapidly under aerobic conditions, depending on the formulation used, and the microorganisms required. In case of biodegradable materials in which starch is a component as an additive to a conventional polymer matrix, the polymer in contact with the soil and/or water is attacked by the microbes. The microbes digest the starch, leaving behind a porous, sponge like structure with a high interfacial area, and low structural strength naturally. When the starch component has been depleted, the polymer matrix begins to degrade due to enzymatic attack. Each reaction results in the scission of a molecule, slowly reducing the weight of the matrix until the entire material has been digested.

In another approach of microbial degradation of biodegradable polymers microorganisms are grown for the specific purpose of digesting polymer materials. This is a more intensive process

6.4 Methods of Biodegradation

that ultimately costs more, and circumvents the use of renewable resources as biopolymer feedstock's. The microorganisms under consideration are designed to target and breakdown petroleum based plastics [5]. Although this method reduces the volume of waste, it does not preserve non-renewable resources.

Photodegradable polymers degrade due to the action sunlight. In these cases, polymers are attacked photo chemically, and broken down to small pieces. It is just beginning, further microbial degradation is must for true biodegradation. Polyolefins (a type of petroleum-based conventional plastic) are most susceptible to photodegradation.

Other approaches developing photodegradable biodegradable polymers include, incorporation of additives that accelerate photochemical reactions (*e.g.* benzophenone), modifying the composition of the polymers to include more UV absorbing groups (*e.g.* carbonyl), and synthesizing new polymers with light sensitive groups. An application for such polymers is in the use of disposable mulches and crop frost covers which face degradation due to sun light and microbial action.

Some biodegradable polymer materials experience a rapid dissolution when exposed to particular (chemically based) aqueous solutions. As mentioned earlier, Environmental Polymer's product Depart is soluble in hot water. Once the polymer dissolves, the remaining solution consists of polyvinyl alcohol and glycerol.

Similarly in many photodegradable polymers, full biodegradation of the aqueous solution occurs later, through microbial digestion.

Procter and Gamble has developed a product similar to Depart, named Nodax PBHB. Nodax is alkaline digestible, meaning that exposure to a solution with a high pH causes a rapid structural breakdown of the material. Biopolymer materials which disintegrate upon exposure to aqueous solutions are desirable for the disposal and transport of biohazards and medical wastes. Industrial "washing machines" are designed to dissolve and wash away.

6.4.1 Soil Burial

Soil varies widely from place to place. As a matter of fact, the term 'soil' is collective term for all the soils. The soil environment is affected by several uncontrolled parameters. The temperature, the soil water content, the chemical composition, geographical factors and pH. Collectively they create different environments and as a consequence the microbiology and the biodegradation activity can change from soil to soil and from season to season. However, defining the term composting is less critical in comparison. Hence, composting environment is rather viable.

6.4.2 Composting

Composting method for degradation of biodegradable polymer is one of the natural processes of degradation. During composting degradation of waste occurs due to biological process. In compost due to biological process biodegradable polymer is converted into CO_2, H_2O, inorganic compounds and biomass without leaving any toxic material. Typical compost is shown in Fig. 3.

6.4 Methods of Biodegradation

Fig. 3 Compost.

$$\text{Biodegradable polymer} \xrightarrow[\text{degradation}]{\text{(Biological Process)}} CO_2 + H_2O \text{ Inorg. Comp.} + \text{Biomass}$$

Preparation of Compost: It can be prepared by organic waste. Common feed stocks are cow dung, cattle manure, food processing wastes, municipal leaves, brush and grass chipping, sewage sludge, saw dust and other by products of food processing.

$$\text{Organic waste} \longrightarrow \text{Soil like product humus}$$

Organic Waste is transformed into soil like product called humus in which microorganism such as bacteria and fungi to break the organic waste material and to produce compost, carbon dioxide, water and heat.

$$\text{Raw material (Organic waste)} \xrightarrow[\text{Air (Oxygen)}]{\text{(Micro-organism)}} \text{Soil like product humus}$$

6.5 DEFINITION OF COMPOSTING

6.5.1 According to ASTM Standards[9.10]

Composting as a managed process that controls the biological decomposition and transformation of biodegradable materials into humus like substance called compost. The aerobic mesophilic and thermophilic degradation of organic matter to make compost:, the transformation of biologically decomposable material through a controlled process of bio oxidation that proceeds through mesophilic and thermophilic phases and results in the production of carbon dioxide, water, minerals and stabilized organic matter (compost or humus).

6.5.2 According to ISO Draft Standards[11,14]

"Autothermic and thermophilic biological decomposition of bio waste (organic waste) in the presence of oxygen and controlled conditions by the action of macro and micro organisms in order to produce compost" and compost is defined as "organic soil conditioner obtained by biodegradation of a mixture consisting principally of vegetable residues, occasionally with other organic material and having a limited mineral contents".

6.5.3 According to British Standards[15]

Composting as a process of controlled biological decomposition of biodegradable materials under managed conditions that are predominantly aerobic and that allow the development of thermophilc temperatures as a result of biologically produced heat, in order to achieve compost that is sanitary and stable".

Compost means solid particulate materials that are the result of composting, that have been sanitised and stabilized and that confer beneficial effects when added to soil and/or used in conjunction with plants.

6.5.4 According to US Environmental Protection Agency (US EPA) [16]

Composting means the controlled biological decomposition of organic material in the presence of air to form a humus like material. Controlled method of composting include mechanical mixing and aerating, ventilating the materials by dropping them through a vertical series of aerated chambers, or placing the compost in pile out in the open air and mixing it or turning it periodically.

6.6 FACTORS AFFECTING COMPOSTING PROCESS

There are several parameters which affect the compost preparation.

(*a*) Temperature

(*b*) Moisture

(*c*) Aeration

(*d*) pH

(*e*) Carbon: Nitrogen ratio

(*f*) Optimum biological activities.

Quality of compost depends upon the raw material and factors that affect the progress of the process. For optimum composting operation, conditions mentioned in the Table 6.3 should be satisfied.

Table 6.3: Range of factors affecting the Composting Process

Temperature	40 to 50°C	Above this the biological activity is reduced
pH	4.5 to 9.5	Above this N_2 is lost as NH_3
Moisture	40 to 70%	–
Air	0.5 to 0.8 m³/day/kg of volatile compost	–
Carbon: Nitrogen ratio	35 to 50 : 1	–
Carbon: Phosphorus ratio	100 : 1	–

Application of compost depends upon the quality and quantity of compost. High quantity of compost is being used in agricultural, horticultural, landscaping, home gardening. Medium quality of compost can be used in application such as erosion control, road side landscaping and low quality compost can be used for land filling or for land reclamation work.

6.7 PHASES OF COMPOSTING PROCESS

Generally composting process occurs in three phases.

(*a*) Mesophilic phase

(*b*) Thermophilic phase

(*c*) Maturation phase.

(*a*) **Mesophilic Phase**: In this phase of biodegradation, degradation process starts and easily degradable and soluble compounds of organic matter decomposed by mesophilic fungi and bacteria. These organic matters, such as mono-saccharides, starch and lipids are decomposed and gets fragmented. This phase of composting occurs between 20 to 45°C.

6.7 Phases of composting process

   (*i*) Initially mesophilic bacteria produce organic acid and pH decreased in the range of 5 to 5.5.

   (*ii*) As the exothermic reaction takes place, heat is released and temperature increases spontaneously.

   (*iii*) Degradation of protein occurs, due to which ammonia is released which leads to increase in pH levels rapidly between 8 to 9.

   (*iv*) This phase is continued ranging from few hours to few days.

(*b*) **Thermophilic phase**: This phase occurs when compost temperature ranges between 45°C to 75°C. This phase is responsible for decomposition of polymeric material.

   (*i*) As soon as temperature of compost reaches to a level of 45°C thermophilic bacteria and fungi start working in the form of specific colonies of micro-organisms and accomplish high rate of decomposition of polymeric material.

   (*ii*) At 55°C and above many microorganisms that are human or plant pathogen are destroyed because temperature over about 65°C kill many forms of microbes and limit the rate of decomposition.

   (*iii*) Using aeration and mixing the temperature in the process is maintained below this temperature level.

   (*iv*) After peak heating the pH stabilizes to a neutral level. This phase lasts from few days to several months.

(*c*) **Maturation phase**: By this time the high carbon source polymeric material is exhausted and compost starts cooling and becomes stable.

   (*i*) Mesophilic bacteria and fungi reappear in this phase.

   (*ii*) In this phase species are different from the mesophilic phase.

   (*iii*) Protists and wide range of microorganisms including worms, spiders, millipedes, ants spring tails, centipedes mites appear.

   (*iv*) The biological process is slow but the degradable material is further humified and become mature.

   (*v*) Duration depends upon the composition of organic matter and efficiency of the process determined by oxygen consumption.

The composting environment is a rather homogeneous ecological niche and can be considered as a micro-cosmo. This is due to the fact that compost is the result of an industrial process. Any composting manager, in any latitude, will impose similar conditions to the composting plant, inspite of different engineering regimes, in order to reach the same purpose; a fast conversion of the acidic, fermenting waste into a stabilized, earth smelling, marketable compost. To obtain this result, the right combination of parameters (such as carbon: nitrogen ratio, water content, porosity, ventilation) must be set at the beginning of the process and controlled during the reaction to assure a reliable conversion. These parameters favour the development of microbial population which will display the same activity and will carry on the same functions. Thus the assessment of biodegradability is facilitated by this rather constant, homogeneous, "standardized" environment. The rate and the final level of the biodegradation of a given polymer will not be substantially different from composting plant to another, because in any case a basically similar environment will be assured.

6.8 AN INTERNATIONAL STANDARD RELATED TO COMPOSTING

6.8.1 ASTM Standards Related to Composting

The ASTM standards related to composting are tabulated in Table 6.4.

Table 6.4: ASTM Standard methods of composting

1.	ASTM D 6400 – 04 Standard specification for compostable plastics
2.	ASTM D 6002–96 (2002) standard guide for accessing the compostability of environmentally degradable plastics.
3.	ASTM D 6868 – 03 Standard specification for biodegradable plastic uses as a coating on paper and under compostable substrates.
4.	ASTM D 6094 – 97(2004) Standards guide to access the compostability of environmentally degradable non woven fabric.
5.	ASTM D 6340 – 98 Standard test method for determining aerobic biodegradation of radio labelled plastic materials in an aqueous or compost environment.
6.	ASTM D 6776–02 Standard test method for determining aerobic biodegradation of radio labelled plastic materials in a laboratory – scale simulated landfill environment.
7.	ASTM D 7081–05 Standard specification for non floating biodegradable plastic in the marine environment.
8.	ASTM D 5210–92 (200) Standard method for determining the anaerobic biodegradation of plastic materials in the presence of municipal sewage sludge.
9.	ASTM D 5929–96 (2004) Standard test method for determining biodegradability of materials exposed to municipal solid waste composting conditions by compost respirometry.

Contd...

10.	ASTM D 5338–98 (2003) Test method for determining aerobic biodegradation of plastic materials under controlled composting conditions.
11.	ASTM D 5526–94(2002) Test method for determining anaerobic biodegradations of plastic materials under controlled landfill conditions.
12.	ASTM D 5988–03 Standard test method for determining aerobic biodegradation in soil of plastic materials or residual plastic materials after composting.
13.	ASTM D 5271–02 Standard test method for determining the aerobic biodegradation of plastic materials in an activated/sludge/waste water/ treatment system.
14.	ASTM D 6691–01 Standard test method for determining aerobic biodegradation of plastic in the marine environment by a defined microbial consortium.
15.	ASTM D 5511–02 Standard test method for determining anaerobic biodegradation of plastic materials under high – solids anaerobic-digestion conditions.

6.8.2 ISO Standards Related to Composting

The ISO standards related to composting are tabulated in Table 6.5.

Table 6.5: ISO Standard methods of composting

1.	ISO/DIS 17088 Specification for compostable plastics.
2.	ISO 14021:1999 Environmental labels and declaration - self – declared environmental claims (Type II Environmental Lebelling).
3.	ISO 14851:1999 (ISO 14851:1999/ Cor 1:2005) Determination of the ultimate aerobic bio-degradability of plastic materials in an aqueous medium – method by measuring the oxygen in a closed respirometer.
4.	ISO 14852:1999 Determination of the ultimate aerobic biodegradability of plastic materials in an aqueous medium-method by analysis of evolved carbon dioxide.
5.	ISO 14853:2005 Plastics–Determination of ultimate anaerobic biodegradation of plastic materials in an aqueous system – Method by measurement of bio gas production.
6.	ISO 14855-01: 2005 Determination of ultimate aerobic biodegradability of plastic materials under controlled composting conditions – Method by analysis of evolved carbon dioxide – Part –II: Gravimetric measurement of carbon dioxide evolved in laboratory-scale test.

Contd...

6.8 An International Standard related to Composting

7.	ISO 15985:2004 Plastics – determination of ultimate anaerobic biodegradation and disintegration under high-solids anaerobic-digestion conditions-method by analysis of released biogas.
8.	ISO 16929:2002 Determination of degree of disintegration of plastic materials under defined composting conditions in a pilot-scale test.
9.	ISO 17556:2003 Determination of ultimate aerobic biodegradability in the soil by measuring the oxygen demand in a respirometer or the amount of the carbon dioxide evolved.
10.	ISO 20200:2004 Determination of degree of disintegration of plastic materials under simulated composting conditions in a laboratory-scale test.

6.8.3 EN Standard Related to Composting

The EN standards related to composting are tabulated in Table 6.6.

Table 6.6: EN Standard methods of composting

1.	EN ISO 14851:2004 Determination of the ultimate aerobic biodegradability of plastics material in an aqueous medium-Method by measuring the oxygen demand in a closed respirometer.
2.	EN ISO 14852:2004 Determination of the ultimate aerobic biodegradability of plastics material in an aqueous medium-Method by analysis of evolved carbon dioxide.
3.	EN ISO 14855:2004 Determination of the ultimate aerobic biodegradability and disintegration of plastics material under controlled composting conditions – Method by analysis of evolved carbon dioxide.
4.	EN ISO 17556:2004 Determination of the ultimate aerobic biodegradability in soil by measuring the oxygen demand in respirometer or amount of carbon dioxide evolved.
5.	EN ISO 20200:2005 Determination of degree of disintegration of plastic material under simulated composting condition in a laboratory scale-test.
6.	EN 14045:2003 Packaging – evaluation of disintegration of packaging material in practical oriented test under defined composting conditions.
7.	EN 14046:2003 Packaging – evaluation of the ultimate aerobic biodegradability of packaging material in controlled composting conditions – Method by analysis of related carbon dioxide.
8.	EN 14806:2005 Preliminary evaluation of the disintegration of packaging material under simulated composting condition in a laboratory scale test.

6.9 METHODS OF BIODEGRADABILITY TESTING BY COMPOSTING

Procedure and principles of some of the important methods detailed above.

6.9.1 ISO 14855-1 and 2

Determination of ultimate aerobic biodegradability of plastic materials under controlled composting conditions – Method by analysis of evolved carbon dioxide.

- Part–I General method
- Part –II: Gravimetric measurement of carbon dioxide evolved in laboratory-scale test.

6.9.1.1 General Method

This method is designed to simulate typical aerobic composting condition for the organic fraction of solid mixed municipal waste.

This is specified method for the determination of ultimate aerobic biodegradability of biodegradable polymer under controlled composting conditions by the measurement of amount of CO_2 evolved and the degree of disintegration of plastic at the end of the test.

- Sample to be examined for biodegradability is mixed with the inoculum and introduced into a static composting vessel, where it is intensively composted under optimum temperature, oxygen and moisture condition for a test period of not exceeding six months.

6.9 Methods of Biodegradability testing by Composting

- During aerobic biodegradation of test material carbon dioxide, water, mineral salt and new microbial cellular constituents are formed.

- **Inoculation:** There are variants of methods for inoculation. Generally vermiculite (mineral bed) is inoculated with thermophilic micro-organism instead of mature compost.

- **Vermiculite:** Vermiculite is clay mineral can be activated by the inoculation with appropriate microbial population and fermentation and used as a solid matrix in place of mature compost in the controlled composting test.

- The vermiculite is represented chemically as $(Mg, Fe, Al)_3(Al, Si)_4O_{10}(OH)_2 \cdot 4H_2O$.

- The advantage of using vermiculite in composting is that neither the biodegradation rate nor the final biodegradation level is affected because CO_2 evolved due to vermiculite is very low and due to which there is no 'priming effect'.

- **Priming effect:** Generally mature compost has large amount of organic matter. This organic matter in the mature compost can undergo polymer induced degradation and the effect is known as priming effect. It affects the measurement of biodegradability directly.

- Test method is designed in such a way, so that percentage conversion of the carbon in the test material with respect to evolved carbon dioxide as well as the rate of conversion could be calculated.

Procedure: Sample is exposed to an inoculum which is derived from compost. Composting takes place in an environment where temperature, aeration and humidity of compost are closely monitored and controlled. The carbon dioxide produced is continuously monitored at regular interval in the test vessel as well as in the blank vessel to determine the total evolution of carbon dioxide (cumulative). The percentage biodegradation is given by the ratio of carbon dioxide produced from the sample to the maximum theoretical amount of carbon dioxide that can be produced from the sample.

$$\text{Percentage of biodegradation} = \frac{\text{(Carbon dioxide produced from sample during composting)}}{\text{(Maximum theoretical amount of carbon dioxide produced)}} \times 100$$

Maximum theoretical amount of carbon dioxide produced is calculated from the measured total organic carbon content (TOC).

Degree of disintegration of sample and Loss of mass of sample are determined at the end of test.

6.9.1.2 Gravimetric Measurement of Carbon Dioxide Evolved In Laboratory-Scale Test

The method which consists of a closed system to capture evolved CO_2 is available to determine the ultimate biodegradability of plastic material under controlled composting conditions in a laboratory scale test.

- The Carbon dioxide evolved from the test method is determined by Gravimetric analysis of carbon dioxide absorbent.
- **CO_2 absorbent:** Absorption column charged with sodalime and soda talc is used.
- The amount of CO_2 evolved is measured on the electronic balance and the CO_2 contents is determined.

6.9 Methods of Biodegradability testing by Composting

- Inoculum is derived from mature compost and inert material such as sea sand.

- Inert material works as a soil texture and acts as an active part of holding body for humidity and micro-organism activity.

- The degradation rate is periodically calculated by measuring increase in the weight of absorption column, which absorbs evolved CO_2 during composting.

- Test is terminated when the plateau phase of biodegradation is attained: standard time for determination is 45 days. But the test could continue for six months at the latest.

Procedure: The sample is mixed with inoculum derived from mature compost and inert material. Weight of evolved CO_2 is measured by increase in the weight of absorption column which is charged with sodalime and soda talc. The degradation rate is periodically measured. The method is designed to yield an optimum degree of biodegradability by adjusting the humidity, aeration ratio and temperature in composting vessel. The percentage of biodegradation is determined by comparing an amount of CO_2 evolved during composting with the theoretical amount of CO_2.

6.9.1.3 ISO 14852: 1999 Determination of Ultimate Aerobic Biodegradability of Plastic in an Aqueous Medium-method by Analysis of Evolved Carbon Dioxide

This standard method measures the carbon dioxide evolved and determines the degree of aerobic biodegradability of plastic material including those containing additives.

The sample is exposed in a synthetic material under laboratory condition to inoculums from activated sludge, compost or soil. If an unadapted activated sludge is used as the inoculums, the test simulates the biodegradation process which occurs in a natural aqueous environment: if a mixed or pre exposed inoculums is used the method can be used to investigate the potential biodegradability of sample.

The aqueous medium comprises:

- Inorganic medium,
- The organic test material (soul source of carbon and energy) with a concentration range of 100 mg/l to 2000 mg/l of organic carbon and
- Activated sludge or a suspension of active soil or compost.
- The mixture is agitated in a test flask and aerated with CO_2-free air over a period of time depending upon the biodegradation kinetics but not exceeding six months.
- The CO_2 evolved during microbial degradation is determined by suitable analytical method *i.e.* CO_2 evolved is absorbed in sodium hydroxide solution and determined as dissolved inorganic carbon (DIC) or by titration method using barium hydroxide solution.
- CO_2 or DIC analyser or apparatus can be used to determine the biodegradability of the sample through titration method after complete absorption in a basic solution like Sodium Hydroxide or barium hydroxide solution.

Procedure: The sample is exposed in aqueous medium under laboratory condition to an inoculum from activated sludge, compost or soil. CO_2 evolved during degradation is determined by suitable analytical method. The level of biodegradation is determined by ratio of the amount of CO_2 evolved during process with the theoretical amount of CO_2 evolved from the sample. The maximum level of biodegradation is determined from the plateau phase of biodegradation curve.

6.9.1.4 EN ISO 14851:2004 Determination of ultimate Aerobic Biodegradability of Plastic in an Aqueous Medium-method by Measuring the Oxygen Demand in Closed Respirometer

This standard method measures oxygen demand in a close respirometer and determines the degree of aerobic biodegradability of plastic material including those containing additives. The biodegradability of plastic material is determined using aerobic micro-organism in an aqueous system.

The aqueous medium comprises:

- Inorganic medium,
- The organic test material (soul source of carbon and energy) with a concentration range of 100 mg/l to 2000mg/l of organic carbon and
- Activated sludge or a suspension of active soil or compost.

The mixture is stirred in closed flask in a respirometer for a period not exceeding six months.

CO_2 evolved is absorbed in a suitable absorber in the top space of the flask. The consumption of oxygen is calculated by measuring the amount of oxygen required to maintain a constant volume of gas in a respirometery flask, or by measuring the change in volume or pressure or a combination of the two either manually or automatically.

Procedure: Sample is exposed in aqueous medium under laboratory condition to an inoculum from activated sludge, compost or soil. The consumption of oxygen during exposure is measured and the level of biodegradation is determined by the ratio of consumption of oxygen in the process with the theoretical amount of oxygen consumption. Nitrification can also occur in the process and can lead to consumption of oxygen, correction for the same has to be done. The maximum level of biodegradation is determined from the plateau phase of biodegradation curve. Carbon balancing should be done to improve the results.

6.10 METHODS OF BIODEGRADATION TESTING

Biodegradation not only depend on the chemistry of polymer, but also on the presence of the biological systems involved in the process.

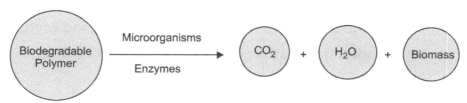

While investigating the biodegradability of a material, environmental affects can't be neglected.

The microbial activity and biodegradation are influenced by:

- The presence of micro-organisms
- The availability of oxygen

- The amount of water availability
- The temperature
- The chemical environment (pH, electrolytes, etc.).

In order to simplify the overall picture, the environment in which biodegradation occurs can be divided in two types:

(*a*) aerobic (with oxygen available)

(*b*) anaerobic (no oxygen available).

These two types in turn can be divided into two different categories

(*i*) aquatic and

(*ii*) high solid environments.

6.11 BIODEGRADATION SUPPORTING ENVIRONMENTS

Different environments in which biodegradation occurs listed in Table 6.7.

Table 6.7: Classification of Different Biodegradation Environments for Polymers

	Aquatic	High Solids
Aerobic	➤ Aerobic waste water treatment plants ➤ Surface waters, *e.g.*, lakes and rivers. ➤ Marine environments	➤ Surface soils ➤ Organic waste composting plants ➤ littering
Anaerobic	➤ Anaerobic waste water treatment plants ➤ Rumen of herbivores	➤ Deep sea sediments ➤ Anaerobic sludge ➤ Anaerobic digestion/biogasification ➤ landfill

The high solids environments are the most relevant for measuring biodegradation of polymeric material, since they represent the condition during biological municipal solid waste treatment, such as composting or anaerobic digestion (bio-gasification).

6.12 DURATION OF BIO-DEGRADATION AND RISK OF THEIR ENTRY INTO NATURE

The main advantage of biodegradable polymers is their less durability, the logical question is; How long does biodegradation take? Principally, it can be presumed that any organic substance under combined influence of the environment and microorganisms will decompose both mechanically and chemically in some time. In terms of the potential spread of degradation products in the environment, and in terms of the usability of polymer products, which need to provide features such as load capacity, water resistance etc.

It is important to understand in what time duration they will decompose and mineralise. Knowing the biodegradation rate also affects the way the material is treated when it becomes waste. Proper waste management of biodegradable polymer is constituted by aerobic or anaerobic decomposition.

In the process of aerobic degradation (with the presence of air), the organic substance - with the help of aerobic microorganisms - is converted into CO_2, water and cell biomass (compost);

In the process of anaerobic degradation (in the absence of air), the organic substance – with the help of anaerobic microorganisms

- is converted into CH_4 and CO_2 (biogas), traces of H_2 and H_2S, and cell biomass.

The rate of biodegradation is therefore very important for polymers that are suitable for composting. It is important to understand here that only some biodegradable polymers are suitable for composting on a domestic compost heap (along with food scraps and other household waste of organic origin).

6.13 BIODEGRADATION MEASUREMENT PROCEDURES

Measurements are usually based on one of the four approaches mentioned here. Before choosing the biodegradation mechanism, it is critical to consider the closeness of fit that the mechanism, will provide between substrate, micro-organisms or enzymes, and the application or environment in which biodegradation should take place [17]. Considering the fact that exposure of polymers, would be in the natural environment, it is considered practically highly important to test the polymer by the field trial and hence soil burial and composting mechanisms are the only options. Exposure in the natural environments provide the best true measure of the environmental fate of a polymer, because these tests include a diversity of organisms and achieve a desirable natural closeness of fit between the substrate, microbial agent and the environment studied, which is likely to differ substantially from many other environments.

6.14 REFERENCES

[1] G. Scott, In Degradable Polymers Principles and Applications, (Ed. G. Scott and Dan Gilead) *Chapman and Hall*, London, Ch. 9:1995.

[2] G. Scott and D. Gilead, In Developments in Polymer Stabilization, Vol. 5, (Ed. Scott . G) *Appied Science*, London 182;71:1995.

[3] D. Gilead In Degradable Polymers Principles and Applications, (Ed. G. Scott and Dan Gilead, *Chapman and Hall*, London, Ch. 10:1995.

[4] G.J.L. Griffin, US Pat. 40,16,117 (1997).

[5] Swift G., *Polym Degr Stab*, 59; 19–24:1998.

[6] K.G. Raj, B.V. Kokta and C. Daneault, *Int J Poly Mater* , 2; 86:1992.

[7] Matsurama S, Mechanism of Biodegradation in biodegradable polymer for Industrial Application, *CRC Press*, Ed. Smith R, Woodhead Publishing Ltd., Cambridge, UK, p-357–409.

[8] Anthony L. Andrady, JMS-Rev. Macromol Chem Phys, C34(1); 25–76:1994.

[9] ASTM D 6400–04 Standard specification for compostable plastics.

[10] ASTM D 6002–96 (RE-APPROVED 2002) Standard guide for accessing the compostability of environmentally degradable plastic.

[11] ISO/DIS 1788 Specification for compostable plastics.

[12] ISO 14851–1: 2005 Determination of ultimate aerobic biodegradability of plastic materials under controlled composting conditions-Methods by analysis of evolved carbon dioxide Part-I: General method.

6.14 References

[13] ISO 14851-2: Determination of ultimate aerobic biodegradability of plastic materials under controlled composting conditions- Methods by analysis of evolved carbon dioxide Part-II: Gravimetric measurement of carbon dioxide evolved in a laboratory skill test.

[14] ISO 20200:2004 – Plastics – Determination of degree of disintegration of plastic materials under simulated composting conditions in a laboratory scale test

[15] British standards institution's – publically available specification for composted materials"(PAS100):www.compost.org.uk

[16] www.epa.gov

[17] J.M. Mayer and D. L. Kaplan in Biodegradable polymers and packaging, Eds., C. Ching, D. L. Kaplan and E.L. Thomas, *Technomic Publishing Co. Inc, Lancaster*, PA, USA, 233:1993.

❏❏❏

Glossary

Aerobic degradation – Biological decomposition in the presence of oxygen or air, where carbon is converted into carbon dioxide and biomass.

Anaerobic degradation – Biological decomposition in the absence of oxygen or air, where carbon is converted into methane and biomass.

Amorphous Polymer – Polymer material that have no definite order or crystallinity. The polymer molecules are arranged in completely random fashion. Ex.PVC, PMMA.

Activated sludge – Biomass produced in the aerobic treatment of waste water by the growth of bacteria and others microorganism in the presence of dissolved oxygen.

Activated vermiculite – Vermiculite colonised by an active microbial population during a preliminary growth phrase.

Biological degradation (bio degradation) – Degradation under the influence of biological systems.

Biomass (renewable resources) – Substances of biological origin, with the exception of those in geological formations and fossilised organic substances.

Bioplastics – Plastics, that are biodegradable and/or based on biomass. In medicine also use which means biocompatibility – compatibility of plastics with human or animal tissues is possible.

Biopolymer – Is a polymer, obtained from either renewable resources and/or is biodegradable.

Biodegradable plastics – Plastics, that are, in relation to conditions of the process, aerobic or anaerobic, entirely degradable into carbon dioxide, methane, water, biomass and inorganic materials.

Certificate – A written statement, issued by an authorised organisation, which confirms that the material or product conforms to the standard. The certificate includes permission to use the certification mark (logo), which informs the users about the adequacy standard.

Compost – Organic soil conditioner obtained by biodegradation of a mixture consisting principally of vegetable residue, occasionally with other organic material and having a limited mineral content.

Composting – The process of organic waste treatment, in which aerobic microorganisms biologically decompose organic material.

Compostable plastics (Plastics suitable for composting) – Plastics, that are (under composting conditions) biodegradable at a rate, comparable to the compost cycle, and that meets the requirements of the corresponding standards.

Crystallinity – A state of molecular structure in some resin attributed to the existence of solid crystals with a definite geometry form.

Degradable Plastic – Plastic designed to undergo significant change in its chemical structure under specific environmental condition, resulting a loss of some properties that may be measured by standard test method appropriate to plastic and the application in a period of time that determines its classification.

Disintegration – The physical breakdown of a material into very small fragments.

Digested sludge – Mixture of settled sewage and activated sludge which has been incubated in an anaerobic digester at about 35°C to reduce the biomass and odour and to improve the dewater ability of the sludge. Digested sludge contains an association of anaerobic fermentation and methanogenic bacteria producing carbon dioxide and methane.

Dissolved inorganic carbon (DIC) (ISO14852) – That part of inorganic carbon in water that can not be removed by specific phase separation for example by centrifugation at 40000 m s^{-2} for 15 minutes or by membrane filtration using membranes with pores of 0.2 with pores of 0.2 μm to 0.45 μm diameter.

Dissolved organic carbon (DOC) (ISO14853) – That part of organic carbon in water that can not be removed by specific phase separation for example by centrifugation at 40000 m s^{-2} for 15 minutes or by membrane filtration using membranes with pores of 0.2 with pores of 0.2 μm to 0.45 μm diameter.

Elongation: The increase in length of a test specimen produced by a tensile load. Higher elongation indicates higher ductility.

Fragmentation – physical (mechanical) degradation of substances/ material into smaller parts.

Graft macromolecule – A macromolecule with one or more species of block connected to the main chain as side chains, these side chains having constitutional or configurational features that differ from those in the main chain. (IUPAC)

Hydrolysis – a chemical reaction, in which compounds react with water molecules and split into smaller parts.

HDPE (High Density Polyethylene) – high density polyethylene.

Impact Test – A method of determining the behaviour of material subjected to shock loading in bending or tension. The quantity is usually is measured is the energy absorbed in fracturing the specimen in single blow.

Impact Strength: Energy required for fracturing a specimen subjected to shock loading.

LDPE (Low-density polyethylene) – low-density polyethylene

Main chain / backbone – That chain to which all other chains (long or short or both) may be regarded as being pendant; where two or more chains could equally well be considered to be the main chain, that one is selected which leads to the simplest geometrical representation of the molecule. (IUPAC)

Melt Index Test – Melt Index Test measures the ratio of extrusion of thermoplastics materials through an orifice of specific length and diameter under prescribed conditions of temperature and pressure. Melt index values is reported in grams per 10 minutes for specific conditions.

Mineralisation – The process of conversion of organic carbon into inorganic forms (CO_2), that occurs under the influence of metabolism of microorganisms.

Oxidation – A chemical reaction (*e.g.* burning, corrosion); a substance that becomes oxidised, emits electrons; in this process it can, *e.g.* merge with oxygen or it emits hydrogen.

PE (Polyethylene) – A plastic polymer with a wide range of applications.

Persistent organic pollutants (POPs) – Organic compounds that are resistant to decomposition in the environment through chemical, biological, photolytic processes; *e.g.* pesticides.

Glossary

Pendent group – Side group: an offshoot, neither oligomeric nor polymeric, from a chain. (IUPAC).

Plastics – Material, whose main components are polymers

Plastics based on renewable resources – Plastics that are produced from renewable resources (*e.g.* cellulose, lignin, starch), and not from fossil fuels.

Plateau phase (ISO/DIS 14855 part 2) – Time, measured in days, from the end of the biodegradation phase until the end of the test.

Polymer – A substance with a high molecular mass, made of perennial basic elements.

PP (Polypropylene) – Plastics with a wide range of applications.

PS (Polystyrene) – One of the most commonly used plastic types.

Ring-opening polymerization – A polymerization in which a cyclic monomer yields a monomeric unit which is acyclic or contains fewer cycles than the monomer. If the monomer is polycyclic, opening of one ring is sufficient to classify the reaction as ring-opening polymerization. (IUPAC).

Strain: The change in length per unit of original length, usually expressed in percent.

Stress: The ratio of applied load to the original cross section area expressed in pounds per square inch.

Sustainable development – Development that meets current needs, without jeopardizing the chances of future generations to meet their own needs.

Theoretical amount of evolved Carbon dioxide ($ThCO_2$) – Maximum theoretical amount of carbon dioxide evolved after completely oxidizing a chemical compound, calculated from the molecular formula and expressed as milligrams of carbon dioxide evolved per milligram or gram of test compound.

Theoretical amount of evolved Methane ($ThCH_4$) – Maximum theoretical amount of methane evolved after complete reduction of an organic matter, and calculated from the molecular formula and expressed as milligrams of methane evolved per milligram or gram of test compound.

Theoretical oxygen demand (ThOD) – Maximum theoretical amount of oxygen required to oxidize a chemical compound completely, calculated from the molecular formula and expressed as milligrams of oxygen uptake per milligram or gram of test compound.

Theoretical amount of evolved biogas (Thbiogas) (ISO 14853) – Maximum theoretical amount of biogas (CH_4+CO_2) evolved after complete biodegradation of an organic material under anaerobic condition, calculated from the molecular formula and expressed as milliliters of biogas evolved per milligram of test material under standard condition.

Thermo-plastics – A class of plastics materials(Linear and/or slightly branched polymers) that is capable of being repeatedly softened by heating and hardened by cooling. (HDPE, LDPE, PP, PS, PVC, PET, Cellulose, PLA, PVA, PCL, PHA).

Thermo-set – A class of plastics materials that will undergo a chemical reaction by the action of heat, pressure, catalysis and so on, leading to a relativity infusible, non-reversible state. Triglycerides, Epoxy, curing agents, Phenolics, Bakelite , and Resins.

Total organic Carbon (ISO 14851) – All the carbon present in organic matter which is dissolved or suspended in water.

Thermal Conductivity – The ability of material to conduct heat.

Thermo gravimetric Analysis (TGA) – A testing procedure in which changes in the weight of specimen are recorded as the specimen is progressively heated.

Thermo mechanical analysis (TMA) – A thermal analysis technique consisting of measuring physical expansion and contraction of materials or changes in its modulus and viscosity as function of temperature.

Ultimate anaerobic biodegradation (ISO 14853) – Breakdown of an organic compound by micro-organism in the absence of oxygen to carbon dioxide, water, methane, mineral salt of any other element present (mineralization) plus new biomass.

Ultimate aerobic biodegradation (ISO 14855 part 2) – Breakdown of an organic compound by micro-organism in the presence of oxygen to carbon dioxide, water, methane, mineral salt of any other element present (mineralization) plus new biomass.

Water absorption – The amount of water absorbed by plastics samples when immersed in water for stipulated period of time.

Index

A

Additives 33
Aerobic biodegradation 179
Anaerobic biodegradation 179
Animal resources 41
Application 16
ASTM Standards Related to Composting 193

B

Bacterial resources 42
Bacterial/Microbial Polyesters 67
Bastioli 91
Bayer 35
Biodegradable and non-renewable 31
Biodegradable and Renewable 32
Biodegradable Plastic/Polymer 7
Biodegradation 6, 177
Biodegradation Measurement Procedures 205
Biodegradation supporting environments 203
Blend Analysis 151

C

Carbonyl Index Study 171
Categories of Polymers 31
Cellulose 62
Chitin and Chitosan 64
Collagen and Gelatin 67
Composition of Starch 54
Composting 186
Cradle to grave 3
Crystallinity Study 150
Crystallization Study 149
Crystallographic Study 160
Curing Study 151

D

Definition of composting 188

Degradable and non-renewable 31

Degradation 6

Degradation Process of Biodegradable Polymer 178

Degradation Study 170

Derivative Thermograviemetry (DTG) Study 154

Differential Scanning Calorimetery 144

Dynamic Mechanical Analyser (DMA) 156

E

Ecoflex 35

EN ISO 14851:2004 201

EN Standard Related to Composting 195

Ethylene- Vinyl Alcohol Copolymer 79

evanescent wave 142

F

Factors affecting Composting Process 189

Fourier Transform Infrared: Attenuated Total Reflectance 139

G

Glass Transition Temperature Study 149

Gravimetric Measurement of Carbon Dioxide Evolved In Laboratory-Scale Test 198

I

Inoculation 197

International Standard Methods for Bio-degradability Testing 181

ISO 14852: 1999 199

ISO 14855-1 and 2 196

ISO Standards Related to Composting 194

K

Karlsson 103

M

Manufacturers 13

Market Areas 16

Maturation phase 190

Mechanical Properties After Degradation 171

Melting Study 148

Mesophilic phase 190

Methods of Biodegradation Testing 202

Index

Milk Proteins 67
Mixed resources based polymers 43
Modifications of Starch 58
Mohanty 108
Morphology Analysis 163
Morphology Study After Degradation 172
Municipal Solid Waste 22

N

Natural Biodegradable Polymers 44
Non degradable and non-renewable 31
Non renewable resources based 42

O

Ok Compost 15
Olefins Vinyl Derivatives 78
Olefins with Acrylate Derivatives 80

P

Petrochemical resources 34
petroleum to potatoes 3
Phases of composting process 190
PHBV 76
Poly (Glycolic Acid) 68
Poly (Vinyl Alcohol) 72
Poly(Butylenes Succinate-Co-Butylene Adipate 78
Poly(Lactic Acid) 69
Polycaprolactone 68
Polyesters 47
Polyethylene oxide (PEO) 77
Polyglycollic acid 5
Polysaccharides 45
Polysaccharides 53
Polyvinyl alcohol 36
Preparation of Compost 187
Priming effect 197
Process Compatibilities 11
Properties of Starch 58
Protein- Based 65
Proteins 46

R

Recycling 10
Renewable resources based 40

S

Scanning Electron Microscopy 163
Soil Burial 186
Starch 53
Starch modified by cholesterol 61

Synthetic Biodegradable Polymers 47

Synthetic Polyester 68

T

Tear Strength 169

Tensile Strength 167

Thermal Analysis 144

Thermo Gravimetric Analysis 151

Thermo Mechancial Analyser (TMA) 155

Thermophilic phase 190

Thermoplastic 39

Thermosets 39

Tone Polymer 35

V

Vegetable resources 40

Vermiculite 197

W

Water Absorption Analysis 167

Weight Loss Study 170

Wide Angle X-ray Diffraction (WAXD) 160

X

X-Ray Diffraction 160